SAIL SOUTH TO THE SUN

Dedication

I dedicate this book to my wife Judy and our family of Simon and Victoria. Their support on and off the water has never wavered, in spite of Father's sometimes unreasonable enthusiasm for his sport of sailing.

Sail South to the Sun

Preparing yourself and your boat
for cruising in warm waters

Clifford A. Stillwell

WATERLINE

Published by Waterline Books
an imprint of Airlife Publishing Ltd
101 Longden Rd, Shrewsbury, England

ISBN 1 85310 375 6

A Sheerstrake production.

A CIP catalogue record of this book
is available from the British Library

Printed and bound in Great Britain by
Butler & Tanner Ltd, Frome and London

Acknowledgements

I wish to thank my secretary June who managed to decipher my
copy and put it down in reasonable order. My thanks also go to
Peter Coles of my publisher Waterline, who has put in many
hours reading and checking the manuscript.

Contents

Preface

Sail South to the Sun

This preface is the result of a rather plaintive joke made at about two in the morning, during a Force Ten Storm some two hundred miles to the North of Gran Canaria. ' I'll write the preface - I've read the book and I'm sailing South to the sun!' I said, as we slid down the face of yet another twenty foot wave in total and utter blackness.

Clifford Stillwell was kind enough to offer me a berth on his schooner *Mishka* for the trip he was making between Gibraltar and the Canaries in order to start the ARC a little later in the year. As his editor, I thought that this would be a splendid opportunity to carry out some research into the validity of what you are about to read. *Sailing South to the Sun* sounded just my cup of tea and having just completed the preparation of Clifford's manuscript, I felt *Mishka* was already an old friend.

You will find within the following pages that the author has some very definite ideas on what constitutes a seaworthy boat. Also, that the blue-water passages that are involved in reaching the warmer latitudes require careful preparation and attention to detail. On the particular night of the aforesaid joke, I was damned pleased that I was sailing with an author who practised what he preached!

During our six day passage, we experienced a Force Eight on the nose through the Gibraltar Strait, a Force Nine to Ten which blew for twenty four hours on our third day and just in case we were too complacent homing in on Gran Canaria - a further Force Ten blast which mercifully only lasted eight hours. That little lot had not created the soporific environment that I had been anticipating - but it demonstrated without doubt that Clifford had got the boat and himself superbly prepared for every eventuality.

I suggest that all the hard won experience and advice that fills the following pages is of great value - if not essential - to any yachtsman contemplating a change from the cool waters of the higher latitudes to the more pleasant warmth of Mediterranean and Caribbean climes.

Peter Coles

The author's current yacht *Mishka* temporarily at rest in a
typically beautiful Mediterranean anchorage.

Introduction

Why not sail your boat to the sun?

Many weekend yachtsmen dream about cruising their boats to far away places, usually where the sun shines all day and the breezes are moderate and warm.

Unfortunately for most, these dreams will never become reality, for the daily grind of managing a job and supporting a family naturally takes precedence over the hobby of sailing.

There are others who, through frustration with their current lives or changes in their circumstances decide to change it all and 'sell up and go' for a few years. These are long term yachtsmen with a new life that is centred around cruising full time and as such are not really addressed by this book.

My concern is for another group who are perhaps bored with sailing their local areas year after year, and yearn to expand their cruising horizons, but are not sure how to go about it, or if it is a practical proposition to do so.

They, like myself, are in all probability happy with their present life styles and do not want to live on their boats full time. They just want a new sailing challenge and to make a blue-water cruise to new countries and new cruising grounds, without staying away from their homes longer than a few weeks at a time. Well, it can be a practical thing to do, admittedly involving a considerable degree of work and planning, but giving an equal amount of fun and satisfaction in return.

Over the past decade, yachting has become increasingly popular in many countries, where the facilities have in general, expanded to meet this new demand. Nowadays it really is not a great problem to find safe harbours to leave one's boat in. In the Mediterranean flights are numerous and cheap, therefore keeping in contact is straightforward for most chosen destinations. All one needs is the will to go, and the commitment to prepare thoroughly.

To take a small yacht on a deep-sea cruise is not difficult, but I believe it is something that should not be undertaken lightly. It is true that some successful cruises are concluded by people with a cavalier attitude towards the sea, sometimes in unsuitable or ill prepared boats, but this is not the sensible way.

Accidents can happen on the best run ships, and it's just common sense to reduce these odds by thorough preparation of the boat and her crew. It will result in making the cruise more relaxing and enjoyable, which is the real reason for undertaking the cruise in the first place.

In the chapters in Part 2, I have set out to examine the major considerations that relate to the choice of a boat and her equipment that will be suitable for blue-water cruising. It is not a finite recommendation in any sense, for all boats are a compromise, and all owners individuals with their own ideas.

I have written broad guide lines, based on 40 years experience as a skipper and owner, in the hope these recommendations will be of interest and assistance to anyone contemplating a blue-water adventure.

PART 1

Chapter 1
The Decision

The steady drum of rain on the coach roof and the low moan of wind in the rigging brought me into semi wakefulness. I gently swam up through consciousness from my deep sleep. 'God it's raining again' I thought as I slowly turned in my bunk to look at the cabin clock, it was 5.40 a.m!

The early morning light filtered through the cabin windows showing heavy leaden clouds racing through a slate grey sky, a lovely summers day is just beginning, what a life! My eyes roamed around the saloon taking in the normal chaos that is so much a part of cruising in a small yacht in England. I was thankful to see that Judy was fast asleep in her bunk across the cabin. I took in the heaps of clothing discarded the night before, some drying out from the wet row back from the town, the boxes of provisions neatly stacked in the starboard pilot berth and the damp oilskins hanging in their locker by the companion-way.

A small wet nose nudged my left ear, it was Tigger our Jack Russell who lived in the port pilot berth, asking me to wake up and take her ashore, her two bright eyes fixed on mine in anticipation. 'Better not encourage her' I thought, with luck she might settle down again and give us another couple of hours rest.

Our two adult children, Simon and Victoria, were still fast asleep in the forecabin, which they shared – not an ideal situation at their ages but one that is inevitable with a wooden boat of only 37 feet overall that was designed and built in the '60s.

The cabin had a slight air of depression, perhaps the inevitable result of four days of unremitting rain. We were on our annual fortnight's cruise and had reached Salcombe in Devon after days of pretty awful weather. A strong head-wind made the beat from the Solent rather long and wet. We enjoyed the two good days in Dartmouth, then a rush to Salcombe between depressions to end here locked into the 'bag' for four days waiting for a weather window to get to the Scillies , our projected destination. This year we were happy to have Robin and Chantale and their young family with us. They owned a 30ft Jeanneau and we were cruising in company. Their boat *Jessica*, was rafted up alongside which meant we had company to commiserate with when things went wrong. To start with, the cruise had gone well with light winds from the North-East, 'perhaps this year things will be different' we said, 'we may get light winds we said', 'even sunshine we said', so much for the optimistic yachtsman who cruise regardless of the weather.

The last four days really had been the limit. We had been bottled up with many other boats sheltering from the heavy winds, and sat watching the surrounding hills disappear at regular intervals behind sweeping curtains of rain and mist. During the brief intervals between the showers we would make forays ashore to buy supplies, walk the dog and splash around on the hills just to get away from the cramped conditions on board. My heart had gone out to the 'girls' they have showed considerable fortitude, in the face of all this bad weather. It's not their hobby, they just come along because their husbands and fathers have this insane pastime called 'sailing'.

As I lay warm and comfortable in my sleeping bag, my mind slipped back in time – just how had all this started? Why had I developed this passion for boats? Why do I put up with all the hard work, expense, agony and discomfort and what do I really get out of it all?

Of course it was all that man Arthur Ransome's fault. As a boy I read and re-read his wonderful books with great intensity. His descriptive prose sowed a seed that grew in my mind. The clean air, the sailing, the freedom and the adventure were all encapsulated within the covers of his books, they set my imagination alight. They stimulated my desires, to get afloat and relive his exploits, it made me determined to get to sea somehow, anyhow, just as soon as I was able.

In reality of course, a fifteen years old boy with very limited finances and living in London had little chance of achieving such ambitions unless a rich Uncle popped up with the wherewithal. No such luck in my case, I had my goal, and if it took sometime to reach it, the rewards would be that much sweeter. In fact it took 3 more years before an opportunity arose that enabled me to actually get my first sail in a real boat.

The school holidays that year coincided with a spell of fine weather. There was much discussion round the breakfast table at home concerning holidays. It took me much of one day of intensive persuasion and lobbying my mother for support, to achieve an agreement to hire a sailing yacht on the Norfolk Broads. My Mother sensibly decided to let the men 'go alone' so it was agreed that Father, my elder brother Peter and I were to venture forth to Norfolk.

As my father and brother had no experience of sailing on boats at all, it was agreed that I, 'the great expert', should make the travel arrangements and choice of boat. I set about it with enthusiasm and complete ignorance!

My choice was to charter a traditional wooden 30ft sloop based on Oulton Broad. A wise decision as it turned out as we needed the open expanse of water to keep from wrecking ourselves and doing too much damage to others. I will draw a discreet curtain over our exploits during that week, but I remember a number of unplanned visits to the reed beds, a few nasty collisions with withies and a rather unfortunate incident at Ackle Bridge! However, with book in left hand and tiller in right, I managed, with help from 'my' crew, to pilot ourselves around the confined waters of Broad's land without doing any really serious damage, other than to my self esteem that is.

It was a hugely wonderful experience for me, although perhaps I can't say the same for my father, who, with luck and a strong sense of self preservation, survived a number of near misses to his head as we gybed all standing round the countless bends in the river. He decided, perhaps with the wisdom that comes with age, that sailing was not really for him. My brother on the other hand felt it did have some good points and in fact sailed with me on many occasions in the future.

The salt water, or perhaps I should say the brackish water of the Broads was well and truly in my veins now and I was to visit Norfolk many times more in adult life.

To get myself a boat, any boat was now the priority in life. In fact it took me two more years before I could fulfil this ambition. I managed through saving hard and going without small luxuries to scrape together £45 to buy my first boat, a National 12ft racing dinghy.

Diana was past her best and getting on in years, but to me she was perfect. I looked past her sprung wooden mast, her split garboards and flaking varnish. To my eyes she was beautiful and she was mine! I set about putting her to rights and soon she looked smart as paint, no matter if she leaked a bit, or that her sails were old and stained. She sailed like a witch, so I raced and day-sailed her with terrific enthusiasm, but with little racing success.

One's first boat is always one's best, no matter what comes later. The sensations, the tentative learning stages, then the thrill when a certain measure of competence arrives are heightened by the years of waiting. That little Uffa King National 12 gave me many hours of the purest delight. She also started me on the long path of learning seamanship and the art of handling a 'tippy' design in heavy seas, baling with one hand and steering with the other, with the crew having to manage the jib and main sheets plus keeping the boat on its feet the best he could; it was heady stuff indeed.

Inevitably, the time arrived when I desired a more competitive design and *Diana* was sold – such is the fickleness of youth! Later my interests lent towards cruising, 'it must be drier' I thought and also we could sleep in her overnight – what bliss. My first 'real' boat was a little wooden 3-ton gaff cutter, built about 1936 by Hillyards of Littlehampton. Again the learning curve was steep, – this was a little ship that carried her way, could not be spun round with a flick of the tiller and had a little jib right on the end of a long bowsprit. This boat was to teach me a great deal, not the least on how to live with a 4 hp two stroke engine. I developed a love hate relationship with that engine. On it's good days I could be heard describing it as a 'character' on the bad days, something else!

Three tons of long-keeled heavy displacement boat with a modest rig was not the most exciting boat for a young man and his friends who had been brought up on racing dinghies. After a couple of seasons I sold her for a smart little Bermudan sloop of five tons. *Jaldi* was built of larch planking on oak by the Lady Bee Shipyard at Shoreham. She had a tucked up transom stern and a small, well proportioned dog-house. Although full keeled she was well cut away forward and promised to be a good performer. All-in-all an attractive little boat but with one disadvantage, her engine was another 4 hp two stroke! 'No, No,' the owner assured me, 'it's a little gem and always

starts hot or cold,' and it must be admitted it did just that when he was around. But as soon as he turned his back and the boat was mine – back to normal! 'It must be a coincidence I thought', blinded by the optimism engendered by a buyer in love with his new boat. Some years later I learnt from a man much wiser than I, the dark secrets of these engines – 'it's very simple,' he said, 'just remove the magneto every autumn at lay up, then keep it in a warm airing cupboard for the winter' – no problem! Apparently it then starts the new season dry and warm and in a contented mood.

Jaldi proved to be a super little boat, but with one vice, she was tender. Her performance in light conditions was perfectly good, but as soon as the wind piped up her windward ability was severely reduced. So much sail needed taking off she lost all sparkle. This was of no real consequence sailing around the Solent, but for extended cruising it was disconcerting and dangerous. I well remember on one occasion when with two friends we planned a cruise to northern France for a break in the summer. Our cruise started well with moderate south-westerly breezes, but as we approached mid Channel the wind started to get up to around Force 6, then we progressively reefed until we reached a stage when we simply could not make to weather. *Jaldi* just lay on her side and stopped, simply sagging away to leeward. Unfortunately my ownership of *Jaldi* coincided with several successive seasons of bad weather, dominated by low pressure systems, giving unsettled and windy conditions. She was not the ideal boat for impecunious young men who wanted to cruise to far away places in the time limited by annual vacations.

During this period we were offered another boat. She had been lying on the hard at Itchenor for sometime and was looking a little neglected. No one knew her parentage for sure, but she was about seven tons and cutter rigged. She was built between the wars of pitch pine on oak. The great attraction to us lay hidden beneath her large cockpit sole, an 8 hp engine! True it was the same make as our other two stroke engines, but this one had twin cylinders and a reputation for starting first go! In a rash moment I made an offer and the deal was done. We immediately made plans to remove the engine and install it in *Jaldi*. 'Just think, if conditions get too bad, we can now simply motor to windward'. The words of wisdom! For of course motoring a small 25ft boat to windward in hard weather is a thankless task at the best of times, and the ability to keep going, butting into head seas is limited in such a small boat. However, nothing daunted the engine was duly removed and we attempted it's installation in *Jaldi*. No matter how we tried, that engine would just not fit, we would accommodate its extra length with a little carpentry, it was its width that defeated us – two inches was all we needed, but it was no use it would not go in – another lesson learned! I now had two boats, one with a dismantled engine and only a single mooring! A quick discussion with George Haines at his boatyard at Itchenor, and we put her up for sale through his brokerage. Much to my surprise and relief, George quickly found a buyer and she changed hands again for the princely sum of £155. The exact price I had paid for *Elvee* some six weeks previously. It seems incredible now, that such a good little boat could fetch so little, but in 1955 this was the going rate. In fact I still have the receipt from Haines to this day. I often wonder what became of her for I was never to meet the new owner.

I had a very experienced crew at this time, two long established friends and my brother Peter. *Jaldi* obviously had her shortcomings and the racing bug was beginning to bite again. It seemed a logical time to move on, find a new boat that would combine cruising ability on one hand, and give us the opportunity to enter a few races, on the other.

I had recently been introduced to Captain John Illingworth R.N. the well known ocean racing skipper, and he was to have a marked influence on my future thinking. Some of my time was in fact spent with the RORC fleet as crew and I felt this style of racing was something I could manage if the right boat was available, and I could find one that would fit my budget.

Illingworth had been involved with forming the Junior Offshore Group over the last couple of years and he was actively campaigning his own boat. I joined the JOG and found a boat that would qualify for these races. Cheaper than racing with the RORC and just as much fun.

I had a long talk with Adlard Coles who was most encouraging about the JOG and suggested I might find an ex 6-metre to convert. In fact he had considered the idea himself, but rejected it in favour of building a new *Cohoe*. But the idea was accepted by us, and we started to hunt for a suitable six. Our travels led us to many places but eventually we settled on a recent 6-metre that had been designed and built by Uffa Fox in 1950.

Noroda was an interesting boat, originally designed with an advanced fin and skeg keel after his Flying Fifteen designs. She had proved too skittish and was gradually modified to a more conventional underwater shape. This gave her good handling but not race winning speed. I bought her on the East Coast after a large dog-house had been fitted, she already had a reputation for fast cruising, as her new owner had just completed a trip to Norway in her.

To meet the JOG rules, we built a watertight cockpit and fitted lifelines. Other than that she was pretty original with her full racing rig, wire backstays and deep draught. We campaigned her as hard as we dare and always seemed to cross the finishing line near the front, but way down the list when handicaps were worked out. To those of you who have not taken a 6-metre offshore, and I am sure there are many, I say avoid it if you can, it's a masochistic exercise.

'Sixes' with their long overhangs, deep, heavy ballasted keels, marginal freeboard, and big rigs take a lot of water over the deck. Now we understood the reasoning behind the hefty dog-house – it was essential to even tolerable living. Windward sailing gained a new meaning for us. No problem with keeping the boat going in strong weather, the problem was to slow it down so that we would not drown, in the cockpit! Those few seasons taught us a lot about boat handling. *Noroda* did not have an effective engine, it was just 3 hp and gave us 3 knots flat out – in a calm. Seamanship is something that can't be learnt from books, it is acquired over time, but we had a great deal of fun and wonderful companionship, we all increased our knowledge even if we were scared to death for much of the time.

I woke from my reverie – still raining, but all was quiet, even Tigger seems to be dozing. I'm certainly very pleased to be here nice and snug in my bunk rather than bashing away out there on some race or the other.

I suppose the real change in my thinking came about when I married Judy. *Noroda* had been sold, and the syndicate split up due to business pressures taking up too much of my time. I had purchased a little Osprey dinghy just to keep my hand in, but after we married and our son Simon was born I replaced it with a small Prout cruising catamaran. This was a weekender design, and I felt it would be ideal to gently introduce my wife, and baby son, to the pleasures of pottering around in a stable boat.

Nothing too adventurous mind, the odd trip to the Isle of Wight and cruising around the confines of Chichester harbour, it seemed just the job. The weather thought otherwise, and the period of ownership was marked by what seemed some of the windiest on record. My problem was to keep this boat from flying with one hull in the air. Below, Simon blissfully unaware of what was going on, would shoot across the cabin in his carry-cot to end up with a crash on the leeward hull. Judy would then let go the jib sheet to rush below. All hell would be let loose above, with me trying to sort out the sails and stop us capsizing – not the ideal introduction to sailing. It was a nice little boat in many ways but far too scary for my young family. I well remember on one occasion in late September we were running up Chichester harbour before a fresh wind, suddenly a violent squall hit us, the water turned white – (as did the crew!) and we took off. The problem facing me was that I was rapidly running out of estuary; ahead was a starboard bend that led into Itchenor reach. To attempt a gybe in that weight of wind was impossible, so I took the only option available, I just kept going straight to shoot up the mud at ten knots and come to rest in the shingle at the top of the bank, safe but thoughtful!

We all make mistakes and that boat was certainly one. After discussing the options with Judy we decided to commission a new cruising catamaran. One that was safe and would be suitable for sailing with a young family. Bob O'Brien was a very popular designer just at this time, so we had one of his designs built. I had the rig modified a little to increase efficiency, and for her type she sailed quite well. She would reach well (as most boats will) and she was very stable. Windward sailing was a little hit and miss, her tacking angles were large and she was apt to make considerable leeway in heavy seas.

Running was comfortable in a waddling sort of way but all in all she suited us quite well. Certainly we used her a great deal and cruised pretty extensively to France and the

West Country. During the three seasons we owned her many miles slid beneath her keels.

I had been brought up on monohulls and really found that multihull sailing was not something I could get enthusiastic over, so started the familiar round of looking for another yacht. This time the parameters were clearly defined, with a growing family (our daughter Victoria was due to be born) we needed a boat that would be suitable for safe family cruising yet be rewarding to sail – *Zola* came into our lives.

The familiar chore of scanning advertisements and contacting brokers started again, we visited many yards and looked over a number of boats without much success.

The trouble was I had a pretty good idea on the sort of yacht that would suit us, and very few, seemed to be on the market just at this moment. Breakfast was an interesting time looking through the mail to see just what had been sent to us. One Saturday morning the details of *Zola* came through, she was just on the market, and was lying at Burnham-on-Crouch in Essex, an old sailing haunt of mine. Her particulars were about right. A 'Breeon' class designed and built by Franz Mass of Holland in 1963. 37ft LOA, 24ft waterline and 5ft draught with a good beam of approximately 10ft 6in. A lovely looking yacht built of good materials and with excellent breeding – this could be it!

A quick drive to Burnham, a short sail and a check on her gear were soon over. An offer was made through the broker subject to survey. This was accepted; the surveyor did his job the next day and as we anticipated found no real problems. 'She's a honey' he said, so the cheque was dispatched and she was ours – all done in ten days – the best ten days work I had ever done!

The purchase of *Zola* also coincided with the birth of our daughter Victoria, so as can be imagined this was a particularly happy time for us.

Zola proved to be all we had hoped for – beautiful to look at, fast and well mannered and surprisingly roomy for her age and type. During the years we owned her we cruised widely and had some wonderful times aboard. I re-engined with the largest diesel I could find that would fit into her, and gave her a new suit of Ratsey sails. Her performance, particularly in light airs was excellent and we frightened a lot of sharp modern boats.

Crash, the main hatch slid back with a bang which brought me back from my dreaming with a start. 'Anyone coming ashore for fresh milk?' It was Robin standing in the hatchway, oilskins streaming and a grin on his face. Tigger leapt past me without a glance, thankful that someone at last was awake and ready to get ashore. The others started to stir; the wind and rain continued unabated and I reluctantly started to dress. 'I must make the decision', I mused, at this rate I shall be too old to do anything about it – then it all came together in my mind. I would make the decision now, we would get away from this cold and rain – we would sail our boat to the sun!

Zola - The author's much loved classic yacht ghosting off Falmouth.

Chapter 2
The Agony and the Ecstasy

Seventeen years is a sailing lifetime. That is the period we owned *Zola*. She had, during this time become a second home for the family. She was in fact very much a personality, a friend who had looked after us, be it in Brittany fogs or Channel gales. Her inherent seaworthiness somehow always saw us through.

Breaking the news to the family that she was to be replaced was a task I shirked for a long time. It is true that on numerous occasions we had discussed the possibility of sailing to warmer waters, but no one, other than myself, ever really believed it was anything other than a 'let's wish' sort of game.

In reality, I felt a little bit like a traitor myself. *Zola* had been such a reliable boat for us, that to contemplate changing her seemed, somehow ungrateful. The reasons for doing so were sound, she was getting a little cramped for our grown family, and I wanted extra room and perhaps a few more creature comforts. She was also unsuitable for a hot climate, in that her construction of strip planking would, I knew, suffer under hot conditions. It would be difficult to adapt her to warm water sailing without considerable cost, sheathing her underwater against gribble and teredo, also putting in refrigeration etc., so little by little I continued the process of self delusion and justification until I was completely convinced, that this was the only sensible course to follow.

Convincing the other members of the family however, was far more difficult, but in the end there was a grudging acceptance that *Zola* was to be sold 'to strangers!' as my shocked daughter exclaimed, and the process of replacement started.

It was fortunate in that Autumn was soon to be upon us. The season was winding down and the family were occupied with other things.

I had decided that for a change I would lay her up ashore under cover in a yard, rather than in the open and under her fitted covers. This was chosen largely to enable potential buyers to view her more easily, without the effort of removing wet covers. It would also enable me to have her checked over really thoroughly to ensure she was in good order before she was put up for sale.

Perhaps it may sound strange, but I was quite determined on two counts. Firstly, that she would be passed on to her new owner in as good a condition as possible, and secondly, I would find her a new owner that would cherish her as we had done. Of all the boats I had owned she was the finest, and I felt that I owed her a debt of gratitude. My responsibility was, I considered, not over simply once I had received my cheque. Emotional nonsense maybe, but boats engender strong feelings of loyalty.

Time is always a consideration in the sales process, but I was in little doubt that *Zola* would sell easily and quickly. She was in excellent shape for her age and I was confident she would survey well, as we had always maintained her fastidiously during our years of ownership.

The problem I faced was to find a suitable new owner. Someone that hopefully enjoyed maintenance, for keeping a classic wooden boat in top order is a time consuming task. I was also very concerned that she would continue to give pleasure to a family, and it would be ideal if I could find someone with young children to buy her.

I decided to price her sensibly and not

hold out for the last penny. I reasoned my potential market would be wider, and give me the opportunity of finding the *right* owner more quickly. Friends thought I had taken leave of my senses, but I persevered and ran some ads in the major yachting magazines. I also prepared a formal portfolio of *Zola* and her history to present to interested parties. The decision to sell her privately, and not through a broker was influenced by the fact that in this way I would be directly involved in the selling process. I would interview and respond to all potential buyers on a personal level, so weeding out unsuitable ones at an early stage.

As I anticipated, a number of enquiries arrived soon after the advertisements appeared, and I was plunged into the time consuming task of showing prospective owners the boat.

This state of affairs continued for sometime, and I received two firm offers subject to survey. I was not very happy so managed to put them off with feeble comments such as 'I will think it over' or 'I have to give someone first refusal'! About two weeks after this I received an urgent call from Cornwall. 'Has she been sold yet? If not we can drive up tomorrow to view her'. I put a few questions and this enquiry seemed to fit into my category of *right* owner very well. – 'Certainly , please come up, I will meet you at the yard and show you over', I said. The following morning a young family arrived, having driven six hours non-stop, they looked most promising, and having owned wooden boats before, knew what they were looking at. They obviously fell for *Zola*, and trying to be as nonchalant as possible mentioned that they thought she was quite nice, and would ring me with a decision the next day after they had discussed her amongst themselves.

Fortunately they did call again with an offer, and it was with much relief that a sale was agreed. They were a charming family of four, including two young children – absolutely ideal from my point of view. The survey was carried out and *Zola* passed with a clean bill of health, so contracts were quickly settled and *Zola* was no longer ours. It is an interesting fact in life, but I feel it is important to trust ones instincts. I knew that this family would cherish my boat, and so it has turned out. Some five years on we still correspond regularly and learn about our old boat's adventures. Kerry and his lovely family receive my grateful thanks for looking after her so well, they are nice people who will, I am sure, adopt a similar attitude to ours, when the time comes to pass *Zola* on.

So much for the agony, the ecstasy was about to begin! – or so I thought.

I had given a great deal of thought as to what our new boat should be, and felt it would be a relatively simple task to find her – really I should have known better!

I drew the parameters carefully. She would have to be substantially bigger than *Zola* to fit all the family and friends, but not so large that Judy and I could not sail her by ourselves. She must be suitable for warm waters, so timber with its high maintenance and vulnerability to hot sun was reluctantly ruled out. She was to be fast under sail, but have an engine powerful enough to give her hull speed in a head sea. Above all, a seaworthy modern design capable of bluewater passages in reasonable comfort.

Not an unreasonable specification to go forward with into the market, I am sure most people would agree. But the road to fulfilment is never straightforward, particularly when it comes to boat purchase, as we were to find out.

Yachting magazines were bought, brokers were contacted, friends were asked, as the search for a suitable boat proceeded. We felt, quite rightly as it turned out, that a standard production yacht was not for us. The only ones that really appealed were the semi custom yachts and these proved to be outside our budget. I looked into the possibility of building new, but the time span of up to eighteen months to launch was daunting, as was the real costs of such an exercise. Our search therefore concentrated on the second-hand market and many miles were driven in search of our ideal yacht.

Soon it was late September and we were no nearer our goal. We were now considering new build again, and had started talks with a South Coast yard on the possibility of building a stock design, but with some individual modifications, when a boat I had always admired, suddenly came onto the market. It was a Surprise 45, designed by the brilliant Dutch navel architect Pieter Beeldsnijder.

The prototype was extensively covered in the yachting press of 1979 and the innovative concept of this yacht caught the headlines. I well remember that his ideas of a modern fast boat that should be 'comfortable' even 'luxurious' rocked the yachting establishment at the time. It is interesting to note that many of his conceptual ideas that were considered revolutionary at the time, such as aft bathing platform, circular saloon seating, twin aft cabins etc. are now commonplace, even on production designs. The rig he drew of an equal masted schooner was very different, but his concept has proven correct as it produces a powerful balanced rig that can be easily managed.

The price for this particular second-hand yacht was high, but we rushed down to the Hamble to look her over.

Simon and I spent the best part of a day on board and her specification was all that I desired, she obviously was the design that fitted our brief and we immediately made an offer.

The broker promised to contact the owners at once but warned that an offer had been made the previous day. He would get in touch immediately a decision was taken.

Imagine our disappointment next morning when the phone rang and the broker confirmed that she had been sold, we had been pipped to the post by one day!

One important consequence of this experience was to realise that I had been gradually talking myself into the possibility of building an unsuitable boat. So I cancelled further discussions with the yard and concentrated once again on the second-hand market.

All this coincided with the Southampton Boat Show. As normal, we enjoyed a family day at the show examining the new boats and gear on display.

This particular year the weather decided to be kind, it was mild and sunny. Late that afternoon I was sitting with the broker who had handled my unsuccessful offer for the Surprise 45, bemoaning the fact that we had missed this yacht, when he casually remarked 'did I know there was another one on the market in Scotland?' On questioning him further it transpired that this was the first production boat in GRP, and enjoyed the advantages of a lead fin keel. He had no particulars available but gave me the owners telephone number. He also remarked that he had heard she was in need of some 'Tender Loving Care', and 'Why don't I fly up and see her?'.

That evening I called Scotland and verified that the price was within our range, and arranged to meet the owner at Glasgow Airport the next day.

I caught the first shuttle out of Heathrow the following morning and was confronted with *Mishka* lying at a marina on the Clyde. At first sight my heart sank a little. She was rather dirty and her topsides were streaked and scratched. Below she was very damp and musty. Her upholstery was worn and stained, both aft cabins were scruffy with damp patches on the linings. I looked into her bilges and found them half full of water, all in all it was a pretty disheartening picture. I asked the owner if I could spend some time on her alone, then proceeded to carefully examine her in detail.

All drawers and cupboards were removed and emptied, cabin soles were lifted, engine casing unbolted and bunks upended. I then carefully worked from forward to aft, examining every nook and cranny that was visible.

I probed and photographed everywhere and made copious notes for reference at a later date. I then sat down at the chart table and listed all the things I considered needed doing immediately to bring her into good condition. It was a long list!

Undeniably, she was dirty, damp and in need of a thorough refit, but mostly it was cosmetic. Underneath the neglect she showed her class. Her woodwork was superb, stylish in design and beautifully executed. The layout was imaginative and practical and fitted our brief exactly. Excitement started to grow, even though I struggled to be objective, I knew here at last might be our new boat.

Just at this moment there was a clatter and the owner came aboard. He enquired if 'I had gone to sleep' as I had been alone for three hours, and 'How about a quick sail before lunch?' The weather was most uninviting, grey and drizzly with a moderate south westerly blowing across a cold Firth of Clyde, but *Mishka* sailed beautifully, close to the wind and fast.

She was very well balanced and seemed light on the wheel in these moderate conditions. Her engine was noisy and underpowered (another note on the list for replacement) but she handled well, turning within a small circle. Unfortunately her sails were pretty worn (out came the list again) but the masts and rigging looked in good order.

Her equipment level was not very impressive, most of it being outdated or in need of servicing. She had, it appeared enjoyed an active life and had a reputation of being an excellent sea-boat. After completing our sail we docked her, then retired to the local yacht club for a very late lunch.

I had made my decision. Provided I could negotiate a sensible price she would be our next yacht. In most respects she seemed ideal for us, roomy, fast and able, with a split rig to enable short handed sailing if required. She had an exciting air about her with modern good looks and an efficient rig. She promised good sailing in the years ahead.

I will draw a veil over the protracted negotiations that surrounded her purchase. (It was in direct contrast to the sale of *Zola*.) I suppose it is in the nature of things that the vendor always wants the best possible price and the purchaser feels he may be paying too much, but things did get bogged down at one time over some pretty silly details. In the end I just agreed and settled the deal quickly, and at last *Mishka* had new owners.

Now the hard work began, I had booked her into a small yard belonging to Edward Dridge at Emsworth, Hampshire. 'Ego' and his trusty team of ex Bowman Yachts craftsmen have a fine reputation and I was keen for them to restore *Mishka* to her former glory.

As the sale had taken so long to finalize, Christmas was almost upon us and her start schedule was January 1st. I flew to Scotland several more times to oversee the new engine installation and trials, also to arrange for her lay up for her trip south. Originally we were planning to sail her down, but the delays had put a stop to that. The insurance company refused to cover her as the season was so advanced.

An excellent local boat transport company drove her down in three days and she was carefully craned into 'Ego's' centrally heated building, just two days before Christmas.

I had drawn up a schedule of work with the yard and we agreed on broad areas of cost. Firstly she had to be gutted inside and the reason found for her dampness. (The surveyor had given the hull a clean bill of health). This meant all linings, cushions and fabrics throughout the boat had to be replaced. We also wanted the hull to be Awlgripp'd, and the decks and deck hardware overhauled. The previous owner had some internal joinery around the chart table carried out and it looked a mess. I planned to completely strip the starboard side of the saloon away, then rebuild with a new chart table, lockers and armchair in its place.

Whilst all this internal work was going on, Simon and I spent every available minute working on her underwater hull. She had passed survey very well with no sign of osmosis, but I wanted to examine every inch of her to satisfy myself that all was in order. It was a huge task, but we carefully removed all the old anti-fouling down to the original gelcoat and much to our relief found her to be in perfect condition. We planned after her five months in a heated environment to have her epoxied in the spring, using the Awlgrip Hull-Guard system as an additional barrier.

As work progressed, more work was uncovered that needed to be done. Meetings were called, with much sucking of teeth and shaking of heads. The plumbing leaked, (the reason for the dampness?) no doubt the result of cold Scottish winters and the failure of the system to be drained – so that was replaced. So were the waterpumps 'worn out I'm afraid'. The electric bilge pumps went the same way , as did the anchor windlass, the gas cooker, the electric refrigerator, the depth sounder, the radar , chart table instruments, lifebuoys, guard-rails, mattresses, sails and sail covers, hood, batteries and two disgusting loos. At this stage I began to wonder what would be left! But as always there is a bright side. The hull was looking superb after its Awlgrip treatment. Inside the joinery was progressing well and Judy and I were enjoying ourselves choosing the fabrics and soft furnishings. So the refit progressed with

all the traumas associated with such an enormous task.

One Saturday morning in February I arrived at the yard to be greeted with gloomy faces. 'The electrician has found problems and would like a word'! This sounds expensive I thought and true to form it was. 'These new electronics you want installing, I can't make any sense of the existing wiring, it would be best if I took it all out and replaced it. Oh, and another thing, why don't I put in new heavy power leads at the same time, and if you like I could put in a 13 amp ring main.' The story had a familiar sound, so being assured the first cost is always the best cost we had new wiring installed – everywhere!

My original estimates had long since been torn up as had the subsequent ones, costs were soaring but so then were our spirits. The yacht was coming along very nicely now, with her flush decks and low coachroof she looked every inch a blue water cruiser. We had put only the best materials and workmanship into her and only hoped she would live up to our expectations.

As every yacht owner knows too well, a refit never ends, but finally I had to draw a line and say enough is enough. She was to all intents and purposes a new boat, fully equipped now to a sensible offshore standard.

So, one grey windy Saturday morning in May, she was launched into the murky, cold waters of Chichester harbour. Her refit had taken a full five months to complete and had cost a fortune, but at last we had our boat shining and beautiful, and ready we hoped, for anything.

The season after her launch was spent evaluating her and sorting out the bugs. As I expected, her sails were really worn out so we replaced them with a new set from North. We chose a fully battened main and mizzen to give us plenty of drive, which was more than the new 80hp Yanmar managed, until we had a new propeller fitted.

The working up year was most enjoyable and we managed some excellent cruises to France and the West Country. It's always exciting sailing a new boat, she feels quite different from one's other yachts, and behaves in a completely individual way. There is no doubt that size makes for wonderfully comfortable cruising, with her long waterline and large resources of stability, speeds are greater too, with less effort and fuss.

Three years on, as I write this relaxing under the shade of the cockpit Bimini, at anchor in the quiet azure blue water of the Mediterranean, I can reflect with a certain satisfaction on the decisions we made.

We made our mistakes, but thankfully, not as many as we might have done. *Mishka* has lived up to her original design concepts well, she has proved herself to be eminently seaworthy, carrying us through heavy weather on a number of occasions with complete aplomb. Her rig, though expensive to maintain, is a success. Judy and I can manage her ourselves if need be, without help, her inherent balance has been a reassuring comfort to us on many occasions.

Living on board in a hot climate can pose its own problems. The extra hatches we put in have been a real blessing as have the large areas of deck space her design gives us.

In retrospect I don't regret one penny we spent on her quality equipment, only the few instances where we decided to buy second best. The sea is a demanding master. For safety sake and peace of mind only the very best is good enough, not only in terms of equipment but in design and planning as well.

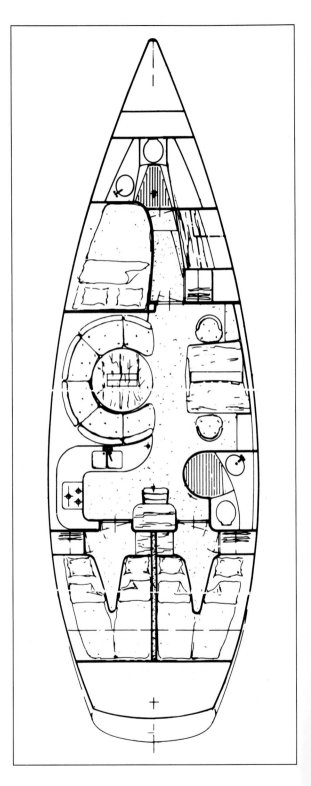

Surprise 45 LOA 45ft Beam 13ft 6in Lead Keel 6 tonnes
             LWL 35ft Draught 6ft 6in Displacement 11 tonnes

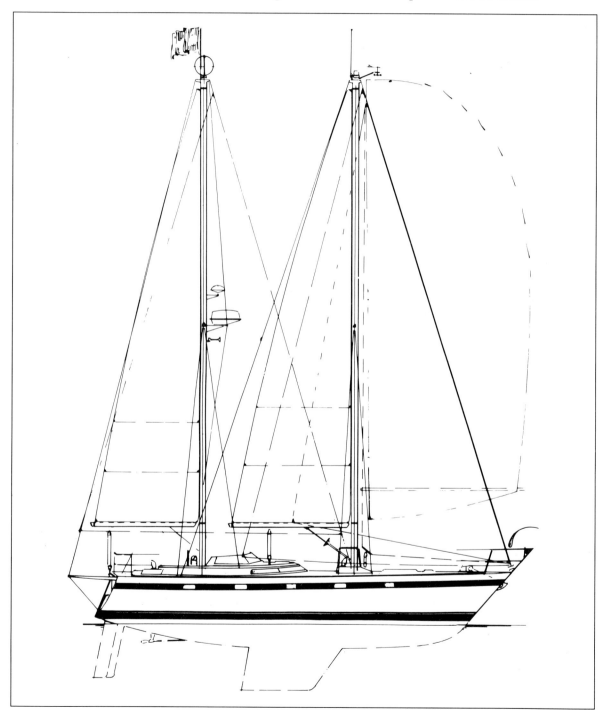

Chapter 3
Sailing to the Sun

The decision to take *Mishka* South had been made. Now the planning began.

I felt we should try to arrange the trip to coincide with the best weather, particularly for crossing the Bay of Biscay. After consulting various weather records, it appeared that May to August were the months with the lowest incidence of gales and provided enough westing was made early on, the weather should be fair. This timing suited my business commitments quite well, so a decision was made to set out on the 31st May the following year. As my birthday was on the 30th we felt this might prove to be a happy coincidence and a good omen.

Mishka was in pretty good shape after her refit and a year long shake down, but I felt a few things needed looking at to make her suitable for Mediterranean sailing. I then sat down and made a list – thirty two separate jobs needed attention before she was in full ocean cruising trim, and ready to go. They varied from simply fitting bolts on the companion-way hatches, to plumbing in a fresh water shower unit in the transom. Never mind, I had seven months before the cruise started – should be plenty of time! Little did I realise that as soon as a few items were crossed off the list, just as many new ones seem to be added. It seemed never ending.

In addition to the fitting out, the actual planning of the trip had to be done. Charts and pilots to be selected and ordered. Lists of spares to be drawn up for all the many electrical and mechanical parts that go to make up a cruising yacht. Menus to be planned and food listed for purchase nearer the departure time. Books to be read and Astro to be learnt, and so it went on. I worked on the boat every spare minute I could manage and gradually she started to come together.

Our route plan was simple but flexible. We proposed to leave Chichester on the 31st May, calling at Falmouth to clear customs and top up with any fresh food and water we needed, then straight across the bay to La Coruña and the Spanish mainland. If we encountered strong southern or south-west winds before clearing Cornwall I proposed to make some westing towards Eire before heading South. If all went well, after La Coruña we would cruise down the Portuguese coast until rounding Cabo de São Vicente, then stay over in Vilamoura for a brief rest. This part of the trip, I estimated to take about eight to ten days. Depending on the weather we encountered we would then sail on to Gibraltar, where two of the crew would fly home. This would leave just Simon and I to take the boat on to somewhere along the southern Spanish Coast, where my wife Judy and daughter Victoria would join us for a leisurely cruise to the Islas Baleares.

I had allowed four weeks in total, which would, I felt, give us ample time, provided of course nothing went disastrously wrong.

Mishka, being thirty five-feet on the waterline is a long-legged boat, so I felt confident we would make good speed if the winds were strong. If light weather prevailed we would call upon the engine, which is capable of giving us a $7^1/_2$ knot cruise speed. I had a few nagging doubts in the back of my mind about fuel cleanliness, as the filters seem to continually block after a few hours running. Also concern over the fresh water tanks. There always seemed to be a little fresh water in the bilges and despite examining all the connections I was unable to trace any direct leak. Other than these two worrying points, the rest of the boat was in good shape.

My birthday party was a happy affair and included the full crew of Simon my son, Peter an ex Commander in the Royal Navy, Robin an old sailing friend and owner of a thirty six foot yacht and myself. We sensibly turned in for an early night for the start at 0600 the next morning. The boat had been fully provisioned the day before. Fuel and water tanks topped up and batteries charged. I planned to leave three hours before high water, then carry the ebb westward leaving enough time to catch the tidal gate at Portland. I knew if I did this there would be no problem at Start Point and we should be able to carry the strong tides round the headlands and buck the weaker tides in the bays. As it happened, this all worked out pretty well for we cast off on time, and waving the girls goodbye headed out to sea in light south-westerly winds and poor visibility.

I was anxious that the start of the trip should go well so was relieved that we did not have a strong beat to windward too early in the trip, so giving us the opportunity of welding together as a complete crew. The forecast in fact was quite good, providing light to moderate winds from the West with fog patches in some areas, and this was just what we got. In fact we sighted no land or ships whatsoever until St Anthony's Head off Falmouth, came up on the bow the next afternoon. The weather had been light but extremely thick, so we motored for about half the total distance. We identified Start Point and the Eddystone on our radar and that was about all. As we motored to the Fal marina the customs zoomed up in their fast launch. They were much interested in us, until they realised we were leaving UK waters rather than returning, but still, it saved us the trouble of seeking them out, and in fact they were most helpful.

A quick turn-round of about half an hour saw us topped up with fresh milk, bread, fuel and water, also newspapers, then we were off again. Whilst we were alongside the jetty we noticed a familiar boat. It was a local Chichester yacht, a racing Swan 43, she was also bound for the Mediterranean and was resting after a traumatic two day trip in fog a couple of days previously. We did not chat for long as I was keen to get moving again, and wanted to make as much distance to the South as possible before a promised low pressure system made its presence felt.

The weather was gloomy and rather damp but our spirits were high. Within two hours or so we had the Lizard in sight and we were going well into a fresh westerly. 'That Swan from Falmouth is on the horizon astern skipper' a voice from behind me said. Robin had been watching for some time as this lovely yacht gradually pulled up on us. Now *Mishka* is no racer, but she is no sluggard either, so we started to concentrate on our sail trim and helming to keep us ahead just as long as possible. The next few hours saw the Swan close up to perhaps one mile from our starboard quarter then gradually come abeam, I could just make out her graceful hull lines as the dusk settled upon us and the wind backed into the South-West and started to blow – the Low had arrived!

Force 8 is not too bad for a boat of *Mishka*'s size, but when it's from ahead and in North Biscay things get lively. Seasickness pills were issued all round as a precaution, and we all ate a good hot supper. *Mishka* was snugged down with ten rolls in the genoa, single reefed main and double reefed mizzen (which as the wind increased later was dropped completely).

We settled into our offshore system of two watches, three hours on and three off, and smashed our way to windward at a steady eight knots. I felt we were a trifle over canvassed for the conditions but the boat

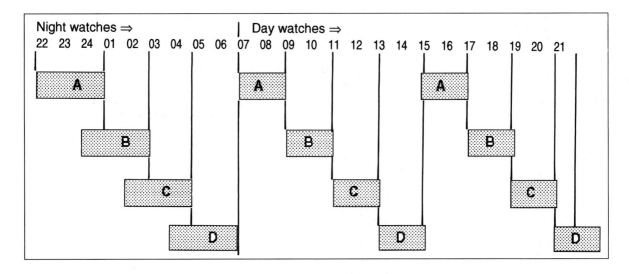

Watch keeping for a crew of four

We have found this an excellent system for moderate weather.

The night watch, although over three hours, is helped by the overlap, so no crew member is alone for more than two hours.

Day watches are two hours in order that the off-watch crew will have enough time to catch up on sleep. The overall rota is deliberately not balanced to give all members a share of the dog watch.

was going beautifully and more importantly we were dropping the Swan!

By dawn the wind had gone down to around thirty knots and showed signs of moderating further, so we gradually increased sail until we were footing along well in breeze of Force 6 or so. There was no sign of our friend so presumably he had overtaken us in the night, or we had dropped him well astern – it depends if you are a pessimist or an optimist. I prefer to think I am a realist, and I think we left him back on the horizon! We never caught sight of him again – perhaps it was just as well!

Day two was pretty average for this part of the world, lumpy sea and moderate to poor visibility. We saw the first school of dolphins of the trip, about ten, leaping and diving all round us. They are truly amazing creatures, friendly, but aloof, doing just what they want and in complete harmony with their environment, yet I feel that humans can almost communicate with them, truly they

have a superior intellect to other sea mammals.

There had been a great deal of shipping about last night, all converging on this north-western corner of France, intent on getting round Brest. As we sailed further south into the Bay, shipping tended to string out, although we were still on the direct route for Finisterre, so a keen look out was necessary. The seas in this area are always big, having no land to stop them, and with the continental shelf to contend with. The short gale had left a big swell and sailing was rather uncomfortable. But still, we could look forward to a good meal tonight! Judy's superb beef casserole, the last of the 'home' food is on the menu. We shall celebrate, and break open a bottle to accompany it. Life still has it moments even in the Bay of Biscay on a dull damp miserable day.

Day three – I took over my first watch of the new day at 2 a.m. to be met with

Gently sailing downwind under poled-out genoa and cruising chute, Simon and Robin enjoy watching the dolphins play in our bow wave.

list prepared by my daughter Victoria, and with a flourish present to the waiting gannets 'Mishka Surprise' – it's a concoction based on tinned chicken, rice, herbs and tomatoes – anyway it tasted fine to me!

We had a different sort of surprise today, the water pressure system went down. After a great deal of huffing and puffing in the bilges we found an electrical connection had corroded on the water pump pressure valve. It was soon replaced but the thought of pumping all that water by hand was not a pleasant one. The other problem with having only a hand pump to the tank is that there would be no hot water from the calorifier, meaning more kettles to boil!

It is quite remarkable how efficient the engine is at producing really hot water. Within five minutes of starting the engine the calorifier is full of steaming water. Six gallons will do the washing and easily give one or two crew members a shower, which reminds me, I must post up the heads cleaning rota, being a cook is one thing but a lavatory attendant really is another!

Day four – The Satnav is giving us trouble again. We have it, and the Decca working side by side and most of the time they agree to a mile or so; but the Satnav just will not input from the heading sensor. I will get Simon to look at it during his next daylight watch. Electronics are a mystery to me. I am quite surprised how well Decca has performed this far south, it seems as chirpy as ever. We have re-programmed the NAVTEX and are picking up the local forecasts very well. The wind is still strong from the North-West and we are running with just mizzen and genoa set. This is a powerful combination for a strong quartering wind with both sails in clear air. I estimate it's Force 7 again but can't be sure as the wind instruments have packed up. One moment they were working and the next, just nothing. The seas are big and very deep blue giving us quite magnificent sailing with dry decks. Good tanning weather too –

penetrating rain driven before a Force 5 north-westerly right into the cockpit. There is just nowhere to shelter and every time the main hatch is opened the rain pours below. Later in the day the weather moderated to a fine drizzle but it is persistent, so there was no let up – wet oilskins every watch!

The wind direction allowed us to run down our rhumb line and as it was still fresh from the North-West we barrelled along at between 6-7 knots with all sail set. Thankfully the rain decided it had given us enough of a soaking to be going on with for one day, and disappeared over the horizon.

Panic! The moment of truth is fast approaching – all my comments about sensible eating, good diets and healthy meals are now to be put to the test. The fresh home prepared food is finished and as volunteer cook for the cruise I am on duty, so I bring out my secret weapon. A recipe the off-watch crew tend to doze on deck in the warm sunshine now rather than stay below reading in their bunks.

Yesterday at dinner we had what I suppose is called a democratic discussion (I knew it was a mistake!) to decide on our route south. Should we stand well off the coast clear of the shipping lanes and separation routes and sail straight down to the Med, or close land, have a day or two in La Coruña, and then sail down the coast route inside the shipping lanes? It was three to one, and I lost – so we press on, well offshore straight to Vilamoura on the south coast of Portugal.

Day five – Light winds are the order of the day with brilliant sunshine. The weather is beginning to get hot now so we rigged the Bimini over the cockpit for some welcome shade. The crew have started to appear in shorts, with lily-white legs and tops, not a pretty sight. We still manage to keep the boat going at about 6 knots with the genoa poled out to windward and the large blooper set flying, with the ever faithful mizzen pulling as well. This is a most successful balanced rig and means the autopilot has very little work to do. Crewing for that matter is a relaxed affair too, we all are able to catch up on some most welcome sleep.

More dolphins visited us, this time about fifteen to twenty, this lot seem to be a much smaller species with a less pronounced bottle nose and darker skins. They stayed with us for about twenty minutes, shooting through our bow wave and leaping right out of the water with sheer exuberance on the large crests from the quartering seas. This is magnificent sailing. As the day wore on the wind increased but thankfully from the right direction. We took in the large chute but continued to run under genoa and mizzen, hard and fast, reeling off the miles.

Day six – A rough night with plenty of wind from the North. The sky is fantastic. I have never seen such brilliance in the heavens. The stars stand out in sharp

Letting the autopilot do the work.
Running hard in fresh conditions we sail into Portuguese waters.

contrast to the jet black sky. I feel I can see right into the depths of the universe. We have been able to watch the satellites tracking across the sky at regular intervals. I also saw a shooting star with its long tail quite clearly – a night for contemplation on the meanings of life if ever there was one.

I often wonder why on some nights phosphorescence is absent, yet on others it's so marked. Tonight it is so strong we can clearly make out the wave crests and the disturbance our rudder is making four feet or so below the surface. We are rolling heavily, so have reduced sail to a reefed genoa poled out on the starboard side.

This last watch before dawn is called the graveyard watch and I understand why. The human body seems at its lowest ebb and this coupled to all the violent motion we feel pretty exhausted; it's difficult to get much rest. I worked out that it takes us a good ten minutes to roll out of the berth and get fully dressed then ready for the watch. Due to the quick rolling and corkscrewing of the boat only one hand can be used to do anything, the other is always occupied just holding on. I am thankful I had all those extra handholds put in, we need every one.

I had just finished dressing when there was a fearful bang from above. The helmsman called out something, but the sound of his voice was carried away by the wind. We tumbled into the cockpit to see the spinnaker boom folded in half lying against the mast and the reefed genoa aback, heaving *Mishka* to. We managed to rescue the two halves of the boom as they had not quite broken apart and lashed them to the chocks on deck. *Mishka* was then gybed back on course and we continued into the night. What had happened was a moments inattention by the helmsman, had brought the boat slightly upwind and this had coincided with a large rolling sea. The sail had gone aback, and the terrific inward pressure of the sail had put too much thrust on to the middle of the boom, and it had simply folded up, another job to be done in Gibraltar! This was our first and only serious breakage so far, inconvenient but not too disastrous. We slowed the boat down a little after this to a more controlled pace as the seas were really quite large, although it could not have been blowing more than about Force 7. It's curious how sometimes the seas are out of all proportion to the wind strength. I hope this is not warning of a heavy blow to come. The NAVTEX forecasts have given us no indication that anything is brewing.

Day seven – The wind moderated overnight and now we are enjoying perfect sailing with a moderate wind and sea. The boat is sailing very well at her most comfortable angle and creaming along at $7^1/_2$ knots with the sheets just started. I have been carefully plotting our course and believe we shall sight land before breakfast. The radar is on its full range setting of twenty four miles but has not shown up Cabo de São Vicente which is surprising, but I know it's there!

In good weather we round Cabo de São Vicente, our first landfall since leaving Falmouth

Great news, I get the off-watch crew on the deck to look at our landfall, the first land we have seen since leaving Falmouth nearly one thousand miles astern. The headland is just on our port bow, standing high and proud from the sea about three miles away. It's a great feeling for us all and we celebrate in the time honoured manner – two slices of toast and an extra cup of coffee. After rounding the Cape we had a long reach of forty or so miles in warm sunshine to Vilamaura. This is a very clean and organized port, and a fitting place to end the first leg of our cruise. The log reading of 1102 miles from Chichester is about right, we have had a fast trip to make Vilamaura in under seven full days. I think we all deserve a good meal ashore after suffering the skippers cooking for nearly a week. Perhaps we shall have an extra day's rest here to get a few of the repairs sorted out, in particular the wind speed and direction electronics, before pressing on to Gibraltar.

The next day at Vilamaura was spent resting and attempting in vain to fix the masthead unit of the wind system. I hope to get it repaired at Gibraltar or if it can't be repaired there, then take it to the UK on my return. (Later the complete 'Brain Box' was found to be faulty and the unit was changed under warranty.)

Running down the southern coast of Spain we were caught by another gale.

We left for Gibraltar the following morning in clear weather, hot, sunny and with a fresh breeze from the South-West. Robin however was giving me a little concern, as he had contracted a minor chest infection, and was not feeling too well. He was getting progressively worse and, as I had put him on antibiotics the day before had hoped to see an improvement. As our 'mast-man' he had been kept very busy sorting out our masthead wind vane problems, and I was feeling a little guilty that no one had taken over this task from him. We began to give him some extra rest, but as we tracked east toward Gibraltar the seas started building again before a strong wind, so that the boat was constantly twisting and rolling, a most trying movement that no one really gets used to.

We are in for a blow without a doubt, a south-easterly in these waters usually means hard weather. It is called a Poiente and when it occurs it's usually pretty vicious. By late afternoon we had a Force 8 on our hands and were running under our normal hard weather rig of reefed genoa and reefed mizzen. The seas were very steep and *Mishka* simply flew. Simon was trying to keep her surfing down the waves at 10-11-12 knots at times. It was a little too much for me so I retired below to get some rest.

Just before darkness fell the masthead wind system suddenly started working again, I really wished it had not, because we were registering Force 9 in the gusts. Not an ideal wind to be blown into the funnel of Gibraltar on. At one time a Spanish frigate steamed up to look us over and presumably to check that we were under control. Lots of waving ensued, then she roared off into the gathering dusk, looking a magnificent sight with spray sweeping over her bridge, and her stern tucked down in the troughs.

By morning we were motoring in an almost flat calm, such is the weather in this area. It's all most disconcerting, but I must admit I was a little fed up with all the hard winds, just as we were about to enter the Mediterranean too – it's supposed to be balmy and hot out here!

Gibraltar was its usual self, busy and slightly care worn, but it's great to be here.

Unfortunately Peter and Robin are due to leave us, and they will be sorely missed. Robin's chest is still not right and we made him promise to visit his doctor on return to the U.K. Simon and I turned our attention to getting the boom repaired and the ship cleaned up for the girls, who were due to join us in three days time. Gibraltar is an excellent place to re-provision and repair a boat. Most repair facilities are available but they are not cheap. Duty-free liquor is a bargain with really low prices. It is well worth the effort to stock up with spirits for a couple of years if you can find the space on board.

Shepherds Marina was, as usual, friendly and an interesting place to stay. There was much coming and going with many fascinating yachts to look at. This place is a natural staging point, with yachts dropping in before their trip south across the Atlantic or east into the Mediterranean. Diesel fuel is reasonable in cost and clean, so it's a good place to top up the tanks.

Soon the boat was ready, and we turned eastward in warm sunshine to sail into the Mediterranean for our rendezvous in two days time. We collected the girls at Malaga airport and started the next stage of our journey into the Mediterranean proper.

I had allowed two weeks of gentle day sailing to reach our final destination on this leg of our cruise, Mallorca. We proposed to use the Balearics as a base for a year before sailing to Greece, Turkey and possibly Yugoslavia. However our plans were flexible and no berth had been booked, if we decided on the day that we should try say, Italy instead of Mallorca, we could do so. Apart from a hard blow the second day when we stayed firmly in port, the thousand miles or so to the Balearics was very pleasant sailing.

Mishka pushed her nose into some lovely ports and anchorages along the southern coast of Spain. The weather was hot and sunny for almost the whole two weeks, and apart from thick fog one day – (a phenomenon that I was amazed to learn is not all that uncommon here) it was uneventful sailing in wonderful tideless conditions. As expected, we had rather a lot of headwind but this was interspersed with some free winds so we managed to sail a good deal of the time. This part of the Mediterranean is bedevilled with a short sharp sea that gets up at a moment's notice. The anchorages and harbours are pretty restless as the swell seems to find its way in everywhere. The steep choppy seas are reminiscent of the Solent under wind against tide conditions and is caused I am reliably informed, by the water being heavier, due to a high salt content. In turn this is the result of the high evaporation rate of this almost landlocked sea. Certainly the boat gets covered in salt crystals, so, after every sail it is common practise to wash ones boat down with fresh water. The crew became relaxed and tanned as the days progressed and finally after visiting nineteen different anchorages and harbours we decided to leave *Mishka* at Cala 'Dor on the east coast of Mallorca. An expensive, but safe marina with good facilities and reasonable access to Palma airport. The log read 2,264 miles from Chichester. We had, on the whole, excellent sailing conditions for the trip, strong winds to get us to Gibraltar then warm friendly winds in the Mediterranean. It had taken us four weeks in total, although we could have done it in ten days less if we had just pressed on with a few stops to refuel and re-provision. But had we done so, we would have missed a host of fascinating experiences.

The only major problems we suffered were a broken spinnaker boom, a split fresh water tank and difficulty keeping our batteries charged, for we relied solely on the engine alternator. The crew were first rate, and my lasting memories of them are, of Robin who would go up the mast at a moments notice, Simon's accurate helmsmanship and generally cheerful demeanour and Peter's incredible night vision, plus an insatiable appetite for banana sandwiches. (Their opinion of the skipper has not been recorded!).

Mishka was completely reliable. She turned out to be fast and beautifully mannered. No heavy water came on deck at all during the cruise, and provided the sail area was reduced and balanced correctly for the prevailing conditions, she was finger light to steer.

The careful planning and preparation played its part, and from the skipper's viewpoint made the cruise, and certainly the hard weather parts, more relaxing than they would otherwise have been. I was sorry to have missed some of the many harbours on the Atlantic coast of Spain and Portugal, but on this trip time was not on our side.

In retrospect, I consider this sort of blue-water cruise is well within the capabilities of any experienced yachtsman and capable crew. The vital link is the boat, it must be meticulously prepared and if need be modified to meet the exacting standards the sea will impose upon it. One may be lucky and have mild weather but if not, it's reassuring to know that one has done everything one can to enable the crew to face difficult conditions in a boat that has been carefully prepared and is fully seaworthy.

PART 2

Chapter 4
The ideal Offshore Yacht

Design & Displacement. The Rig & Sails

I have been asked on numerous occasions 'what makes an ideal offshore cruiser?' and of course there is really no finite answer. So many design considerations influence the behaviour of a yacht at sea, it is an amalgam of many features, that make one yacht stand out from another. What is clear however, is that some modern design and construction techniques have in recent years greatly influenced thinking on seaworthiness. Many, hitherto, long held beliefs have been challenged and we have seen some remarkable voyages completed successfully with what a few decades ago would have been considered undesirable offshore designs.

The 'died in the wool' traditionalists would have us believe that unless a yacht had a full keel with attached rudder, is of heavy displacement, and has the appearance of a Colin Archer design, it is not a real sea boat. This is of course complete nonsense. Modern construction techniques and materials have given us superb, light, strong boats that are supremely seaworthy, and in many cases superior to older heavy designs. High topsides keep a boat dry, light hulls 'give' to the seas more easily and can be driven fast to windward in strong weather, whereas their old fashioned cousins would possibly need to heave-to or lie a hull.

Ideally a yacht should be designed for the purpose the owner will put her to. But of course this is not always possible and compromises have to be made. When contemplating a world cruise the main requirements would centre around comfort in a seaway, generous load carrying capacity and the ability for the yacht to look after herself without the crew constantly having to tend the helm and sails. Such a boat would probably be of moderate displacement with a well balanced hull, and a rig that is easy to manage.

But most yacht owners are 'not selling up and sailing away' so their requirements are for a less dedicated cruising design. They want a boat that will make fast effortless passages of up to several weeks duration at a time. They also want a yacht to be comfortable and spacious to live on in harbour and fun to sail. Many modern designs fit this category perfectly well, some of them standard production cruisers, that just need a little adaptation and re-equipping to turn them into safe blue- water yachts.

Part two of this book will examine the main priorities of design, construction and equipping a cruising yacht for a blue-water trip to the sun.

Design and Displacement

Yacht designers have a great deal to answer for. Theirs is not a precise science as many will admit and many's the time when struggling to control an unbalanced yacht in heavy seas, that I have cursed to myself and wondered just why the particular yacht I was sailing, had been designed the way it was.

Some designers in the past such as Laurent Giles, Robert Clarke and others have left us a legacy of beautifully balanced yachts that are a true delight to sail.

I well remember a race some years ago in a lovely ocean racer called *Uomie*. (She was drawn by another artist, Arthur Robb.) It was a typical heavy weather affair and called for a great deal of fortitude from the crew. I was one of the helmsmen and was completely bewitched by this boat. She had a wonderfully sensitive feel about her, totally responsive, with an almost sensual touch to her helm.

Cruising in her will always be a pleasure for her fortunate owners.

Modern design has of course moved on considerably since those days. It is now more of a science than in the past, many lessons have been learnt and the race-bred yacht has, without doubt, been the catalyst for many new ideas. The contemporary yacht at its best is superbly seaworthy, fast and responsive, with strong light hulls that are easily driven. What perhaps some lack in elegance, they make up for in function beauty.

Displacement

Displacement is the volume of water that a hull displaces when immersed. It also gives a clue to the type of boat she is likely to be.

Many production cruisers these days tend to be light to moderate displacement, which provides for economical building costs, fast hulls and easily managed rigs. For long distance cruising, care should be taken to ensure that the chosen design is not too light, as it will severely limit the amount of stores that can be shipped on board. If you overload a light hull it will result in sluggish performance and possibly reduced seaworthiness.

Heavy Displacement

In general, a heavy displacement yacht has these *advantages*:
1) Steady motion in a seaway.
2) Greater load carrying capacity for a given size of hull.
3) An air of solidarity and seaworthiness.
The *disadvantages* are:
1) Higher cost in building for given size of yacht.
2) Bigger rig and engine to maintain performance.
3) Harder for her crew to work (heavier gear).
4) Can be wetter and less nimble than her lighter sisters.

Moderate to Light Displacement

These yachts have some clear *advantages*:

1) Low building costs relative to size.
2) Smaller, less expensive rigs and engines.
3) Easier to handle.
4) Fast and responsive to sail.

Their *disadvantages* are:
1) Tendency to be lively in a seaway.
2) Not good load carriers for their size.
3) Overloading effects trim and speed

Light Displacement

There is a misconception that light displacement yachts are flimsy, less seaworthy and a handful in fresh down wind conditions. This is just not so; these are matters of design not weight. A well designed light boat is just as seaworthy (and some say more so) as a heavy one. The type suffers from the bad reputation of some I.O.R. racing designs. These yachts have been designed to a restricted rule that favours shallow saucer shaped hulls with low form stability together with vestigial keels and rudders. Their light weight is merely a function of making them fast under sail. I.O.R. racers usually have large experienced crews to keep them on their feet and under control, so the fact that they can be difficult to handle downwind is really of little significance; winning races is their priority. I do feel however that the I.O.R. influence is not a good one in terms of producing wholesome cruising designs, and it is unfortunate that some manufacturers produce cruising boats that are adaptations of these rather extreme designs.

Balance of the hull, especially when heeled is an attribute that should be sought after in a serious cruising boat, for there is nothing more exhausting than trying to steer an unbalanced hull in strong conditions for any length of time. Balance also directly influences seaworthiness, and makes the boat far easier to handle. She will keep going in almost any weather, provided she is correctly canvassed.

In recent years we have seen the development of the ULDB, that is the Ultra Light Displacement Boat. These are high speed flyers that rely on very light construction, large rigs and flat runs to their underbodies. In the smaller sizes these are great fun but, not really suitable for ocean cruising. Their motion is violent and to maintain their speed potential, great care has to be taken not to overload them. In the larger size ranges however, say from fifty feet upwards, they can provide excellent fast cruising, and many American designs have proved to be most successful.

Seaworthiness is not a result of displacement but of design. There are superb ocean cruisers both heavy and light, the choice is rather one of preference of a type and what the boat will be used for. Currently there is a move to increase beam to improve accommodation. This can be carried too far. Fat little boats are great in harbour but at sea can be a real problem. The underwater shape when heeled become unbalanced and they start to gripe up to windward, putting strain on the rig and the crew.

For a serious cruising boat it is far better in my estimation, to sacrifice a little space, slim the hull down to improve weatherly ability and if the extra accommodation is vital, increase the length of the hull. The boat will sail faster, be kinder in a seaway and be much easier on her crew.

I have personally selected a medium to light displacement yacht that is well balanced with good directional ability. She is fun to steer and is dry under most conditions (high topsides also help), but I have to be careful not to overload her. Perhaps the attribute I most value, is her directional control and stability. This is due in some measure to her size and volume, but largely to her tank-tested modern design.

That size equates to comfort at sea there is no doubt. A long waterline gives a softer ride and is of course related to a yacht's displacement speed. Spaciousness, comfort and more luxury accommodation can also be fitted into larger hulls but size does not immediately effect seaworthiness.

All things being equal, the ideal size of yacht for serious cruising is determined firstly, by the number and age of the regular crew, then, by the use she will be put to. I prefer to go for the largest boat I can manage with my regular crew. In our case it's forty five feet. We have come to this conclusion, and it has worked out well so far by choosing a boat with a split rig and powerful engine. There is little doubt that the extra stability a long hull gives is a major factor in enjoying cruising in all weathers. Also the extra space on deck, cockpit and accommodation, contributes to the enjoyment of living on board. Some owners relate the maximum size of mainsail they can handle as the limiting factor, but with modern, powerful in-mast furling rigs, this is no longer a realistic guide. We have found that when short handed, with just my wife and I on board, the real limit in handling the boat is not at sea but in confined marinas.

Berths seem to get more constricted each year and handling a high volume 45ft yacht in confined conditions, with just one propeller is sometimes a matter of concern, particularly in a cross wind. I really feel handling a yacht much over fifty feet by ourselves would be difficult, unless of course we had a bow thruster. Even then there is the physical problem of leaping down from even higher topsides onto the dock to collect lines etc. The other end of the scale must be around 30ft LOA. Boats smaller than this put a lot of strain on the crew in heavy weather. The boat will also be relatively slow and making to windward under strong conditions in such a small yacht can be an exhausting experience. The other factor is living aboard in a warm climate. Small yachts can be pretty airless at the best of times (we call it 'cosy' in the cold northern climate!) but there is no doubt that a little

extra space can mean all the difference in enjoyment when a number of people are living together in a confined area.

The Rig and Sails

There is much debate on what constitutes the best rig for blue-water sailing. The current trend is in favour of the cutter, and there is no doubt this is a handy and powerful rig for most conditions on medium to large yachts. I really don't see the argument for this rig in boats of less than 40ft LOA. It is more complicated than the sloop, needs more gear to handle it and in most conditions is slower than the sloop-rigged yacht. With modern sail handling gear, such as powerful self tailing or powered winches and reefing headsail systems, the sloop has a lot going for it. It is simple, strong and the fastest of all. With just two sails it's also the cheapest to buy and maintain. I would feel confident of handling a yacht up to forty five feet with a sloop rig with just two people, and perhaps a sixty five footer with the back up of powered winches.

The ketch is not particularly fashionable at the moment, I often wonder why. Perhaps it's because in many instances in the past it has been used on motor-sailers, where maximum efficiency under sail was not a serious priority. In my experience, if the rig is designed correctly and the mizzen made a sensibly large area, the rig can be extremely fast and weatherly. The flexibility it gives is also a major bonus for short handed sailing, and it can be made to balance under all conditions. Shorter, lighter masts can also be used to increase stability. The main disadvantage with a ketch is that the mainsail is blanketed by the mizzen when running directly down wind, which incidentally is the one position I avoid if at all possible. This position can be dangerous in heavy weather and is normally pretty uncomfortable. When ocean cruising it can simply be avoided by bearing off a little to keep all sails pulling, the extra distance actually sailed will be very little extra.

We find with *Mishka*'s high-aspect-ratio main and mizzen that she can be sailed as a ketch or sloop. When running with the wind on the quarter in strong conditions the ideal rig is genoa and mizzen. She will then log eight knots plus, in a relaxed manner, with no sail flapping or flogging. When the time comes to reduce sail again (about 35 knots) we drop the mizzen completely and run on boomed-out genoa, reefed to the conditions. This has worked out very well in practise, and there is no doubt that having a roller reefing genoa is vital for ease of handling.

Sensible yachtsmen on long distance hauls plan downwind routes whenever they can, so the rig should be adapted for these conditions if at all possible. To sail a few hundred miles extra is no real hardship if it can be done in comfort, without turning to windward too often.

Offwind Rigs

On short handed yachts sailing longish distances it is sometimes a problem setting enough sail area to maintain good noon-to-noon runs in light conditions downwind. Spinnakers although useful at times are to my mind not really a seaworthy option with short handed yachts, particularly for sailing at night. A cruising chute is really not a great deal better, for to be really efficient under some conditions, it needs the same gear as a spinnaker.

The old style twin poled jibs have the considerable disadvantage that they set only a relatively small area, so are painfully slow in light conditions. They were popular before the advent of wind vane steering and auto pilots because with the addition of quarter blocks, the sheets could be led aft to the tiller or wheel to steer the yacht.

Most modern blue-water cruisers use their existing mainsail, with the large genoa poled out to windward. This is a sensible, easy to handle rig that causes little trouble even in strong weather. I have given the 'ideal downwind rig' a lot of thought and if I were re-rigging or starting afresh with a new boat it would be rigged as a ketch with twin fore stays, one set behind the other. Both genoas would be of maximum size to give perhaps 15% more sail than the normal mainsail/genoa combination. They would both be controlled by roller reefing systems, the aft one would be used for windward work in the normal way and the forward genoa rolled out for running before the wind with mainsail furled. A light pole attached to the mast on a separate track with its own downhaul and topping lift would keep them under control. The sails would have their own sheets led aft to their own self tailing winches, and once set should be little trouble. The large sail area would ensure the boat maintains a good speed in all but the very lightest airs. When the wind piped up it would be a simple matter to roll up both genoas by the required degree, so keeping the sail area in balance, right down to gale force conditions. The initial cost would be high, but not all that more than a spinnaker and its gear. The other plus point to this system is that the mast would be superbly stayed with an extra forestay supporting the mast.

Heavy Weather Rigs

No boat should put to sea on a long cruise without the ability to set storm canvas. This is particularly important in smaller sized yachts that have only moderate stability.

The old argument of being able to sail off a lee shore in any conditions is a bit old hat now, as most yachts have reliable diesel engines, efficient rigs and strong gear. They are also very close-winded in comparison to yachts at the turn of the century.

Yachts can, and do, get overwhelmed in storm conditions, and one only has to experience a really severe gale to understand the enormous forces at work. So some form of storm canvas to my mind is essential. Just as important, is the ability to set it in conditions of extreme weather. There really is little point in having a trysail available if the weather is so bad that no one can venture on deck to put it up!

A trysail is an excellent hard weather sail but to be effective at sea, it must be fitted on its own track on the mast. It should be kept on its track in a bag (with drain holes) and lashed at the foot of the mast ready for hoisting. The sail should also have its own sheets permanently bent on ready for use. Don't use a block at the clew to give extra purchase, these can be very dangerous swinging round one's head. On boats up to forty five feet or so, bend the sheets directly onto the sail with a lashed bowline. The sheets can then be led back to the genoa winches via strong quarter blocks.

There is a temptation to make trysails very small, but try to avoid this, they should, if anything be quite large so that they give plenty of drive, they can then also be used as a hard weather mainsail in moderate gale conditions. Have the sail radial cut and put in a line of reef points. I believe the correct size is approximately thirty percent of the size of the normal mainsail, no smaller. On stiff boats from forty five feet upward this can be increased to thirty five percent but with two lines of reef points.

A storm jib should always be carried, and again, thought should be given to how it can be set efficiently and quickly. Boats with roller reefing headsails are a problem, but don't, as was recommended to me by one eminent designer, fit the jib with parrel beads and hoist it over the rolled sail. If you have a roller headsail system do the job properly and fit a detachable forestay. Make sure the deck fitting is really strong and beyond doubt, with heavy underdeck support. When not in use the spare stay can be lashed securely to the standing rigging out of the way. Fixing the stay down to the deck can be achieved by using a normal

This inner forestay can be set up quickly using an over-centre hook with a bottle screw incorporated in the unit. The deck fitting has substantial under deck reinforcement.

bottle screw, or if a faster method is required, one of the patent over-centre hooks can be used. We have one of these on *Mishka* and it works very well. We also have a permanent fixing eye on deck for the sheet leads, and the storm jib has its own sheets permanently bent on to the sail when stowed in its bag. I work on the principal that in the event of needing a storm sail, it's bound to be very unpleasant and difficult to work on deck. Knowing my luck it will also be at night and raining heavily, so anything I can prepare beforehand, to reduce the possibility of problems and discomfort, must be well worthwhile. The chances are that with careful route planning, storm canvas will not be required very often, but it is a wonderful insurance, and in extreme conditions could be the saving of your boat. Of course it is all very well to theorize sitting at home in the comfort of ones armchair. But in reality one can find oneself in difficult situations at sea when one least expects it. I have found in these circumstances you simply do the best you can and it usually works out well if you and the crew don't panic and think things out carefully before acting. A well found boat is amazingly seaworthy, and can usually take a great deal more punishment than her crew.

Reefing

The old maxim, 'reef early and reef deep' is now out dated advice for yachtsmen cruising far offshore. Nowadays with reliable modern reefing systems, reducing sail is a quick and simple operation in comparison to the situation facing our forebears. Then, with large sail areas of stiff unyielding canvas, large booms that overhung the transoms of their yachts and antiquated points reefing gear, it was no wonder they had to 'reef early and deep'.

With in-mast or in-boom reefing for example it's a simple task to roll in a few feet as the wind picks up. Even with jiffy reefing, provided the system is well designed, reefing is fast and trouble free. On the other hand it's just as important to be able to shake out reefs quickly, for winds rarely stay strong for long. Much of the pleasure of cruising lies in the ability to make fast passages, so being able to get canvas back on again and the boat moving with the minimum of effort, is very important.

Most cruising yachts these days have some sort of roller reefing system for their foresails. The concept is now well proven offshore and it's just common sense to take advantage of this modern aid. The one proviso is that every yacht contemplating a long cruise should have a method of setting a storm jib completely independently of the reefed down genoa. It is an accepted fact that once the roller reefed genoa gets much smaller than, say, working jib size it suffers from excessive bagginess and dramatically looses efficiency to windward. Also as the genoa is progressively reduced the clew rises higher, so necessitating the sheet leads to be moved forward. The end result is a small baggy sail set much too high up the forestay, which in turn reduces the boats ability to work to windward. This, plus greater heeling moment due to the foresail's high position.

What Reefing System to choose?

The ability to reef sails quickly and simply is an asset that will be appreciated by all the crew. It is the nature of all things human, that reefing is usually left late and one is faced with pulling down a reef in uncomfortable conditions on a wildly pitching deck. With the advent of roller systems things are very much better and they are a considerable asset to the cruising yachtsman.

Roller Reefing Foresails

The roller reefing foresail is now standard rig on most cruising yachts. It has been well tested in ocean cruising conditions and can be accepted as pretty well foolproof. There are many models on the market, and as always it seems the most expensive are the most reliable in the long run. On *Mishka* we have the well proven Harken™ gear which seems particularly robust, and to date has given us little trouble. Maintenance is virtually non-existent, we simply give it a good douse of fresh water from time to time to wash the salt away. I check over the swivels at the head and tack each season and squirt them with WD40™, more with the feeling that it must need 'something' doing to it, rather than in the real expectation that I am carrying out essential maintenance. We have on occasion suffered some snarl ups in the winding drum and

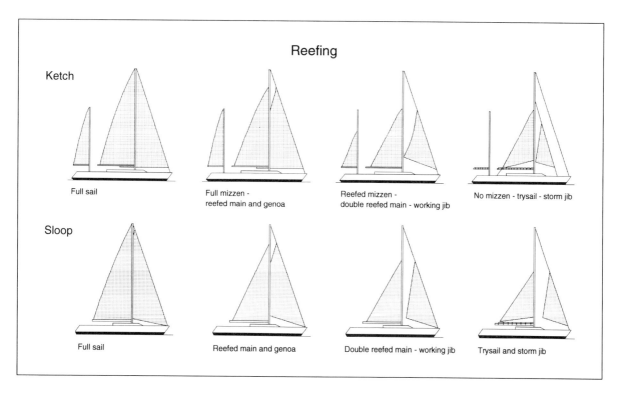

Reefing

Ketch

Full sail

Full mizzen - reefed main and genoa

Reefed mizzen - double reefed main - working jib

No mizzen - trysail - storm jib

Sloop

Full sail

Reefed main and genoa

Double reefed main - working jib

Trysail and storm jib

With ketches there are different sequences for reefing the sails depending on the boat's heading, on or off the wind. Some yachtsmen recommend for strong offwind conditions using just the mizzen and small jib with the mainsail furled.

Sloops have a more limited choice, but cutters can utilize their inner jib after furling the Yankee for windward sailing in fresh conditions.

have altered the lead to ensure it is always fair, and does not foul the lips of the drum in any way, but this was a rare occurrence, and in all probability my fault rather than the manufacturers.

Roller Mainsails

This is the big step forward in mainsail reefing over the last decade. The move started with in-mast systems and is now being extended to in-boom models. There is little doubt that the convenience of these systems is impressive. They have enabled larger yachts to be managed with smaller crews and the husband and wife team can now manage quite a large boat themselves without recourse to extra family or friends.

There is a downside of course, and in the case of mast reefing systems the mainsail has to be cut with no roach to enable the sail to reef within the mast slot. The foot is also cut high to fit the boom track slider. The consequence is that sail area is lost and to my eyes the rigs look a little 'skinny'. If the boat has been designed with this system from the outset the 'lost' sail area can be accounted for with a taller mast or longer boom, but of course the extra weight of these items has to be accounted for. On large yachts the actual rolling sequence can be carried out by electric motors or hydraulic systems so making the whole affair of reefing even more painless.

Reliability has now improved a great deal,

This yacht has been fitted with a bolt-on mast furling system for both main and mizzen. Although this has reduced the sail area a little, the owner considers the simpler handling and reefing characteristics more than compensate for the slight reduction in light airs performance.

and many yachts fitted with in-mast roller systems proceed to cruise extensively with complete reliability.

Roller Booms

This system is now becoming popular and in some cases is preferred to in-mast reefing. The principal is that the mainsail is rolled inside the boom as opposed to outside as in the older roller reefing systems. The advantages are that the sail is kept clean and protected when rolled. The sail is rolled round a rod which has bearings fitted at each end to reduce friction, and the sail can be cut with a normal roach. Some systems

also accommodate fully battened mainsails which offer the best of all worlds. I believe that the system can be used for reefing when the boat is being sailed free, but I have no personal experience to back up this claim. Provided these booms can be built with a long life expectancy and reliability, I see them as an excellent method of reefing. Certainly from the viewpoint of protecting the sailcloth from harmful UV rays they are first class and they also have the advantage over in-mast reefing in that there is much less weight aloft.

Slab or Jiffy Reefing

Slab or Jiffy reefing has come back into fashion and if correctly set up it is extremely quick and effortless. Some yachts have all lines led back to the cockpit so that the complete operation can be undertaken from the shelter of the cockpit. One of the greatest advantages with this system is its simplicity and reliability, assets that should not be underrated when cruising offshore. Some sails are still made with reefing points to enable the bunt of the sail to be gathered up neatly. These are really unnecessary, I much prefer a line of reef eyelets and a bungy line taken to a few hooks on the boom, it is then a simple task to slip the line over the folded sail and under the hooks.

If a decision is taken to take all the reefing lines back to the cockpit, ensure as much friction is removed from the system as possible. Run the lines with the minimum of angles and if blocks are required, use the roller bearing types.

It's worth looking at the actual set up of the reefing cheek blocks on the boom, it is surprising how many are incorrectly led, causing inefficient reefing and stress on the leech and clew of the sail.

The cheek blocks should lead the reefing line pendants two to three inches outboard of a perpendicular dropped from the reef cringle to the boom. This will ensure the correct tension is applied to the reefed sail.

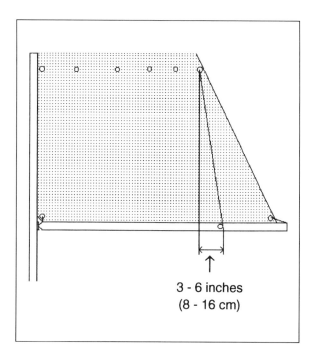

3 - 6 inches
(8 - 16 cm)

Correct offset to ensure tension when reefed.

The best boom systems have a slide or worm gear incorporated within the aft section to allow tension adjustments to the foot of the sail.

Roller Reefing

Although fitted to relatively few yachts nowadays, roller reefing, if set up and maintained properly is an excellent method of reducing sail.

I have found the best way to use the gear is to heave-to, then slacken away the mainsheet until almost all the wind is released, ease away the main halyard, then roll the sail in. If a little wind is left in the sail it seems to keep the leech tight, so no one need attempt to pull it aft as the sail is rolled up, which is a dangerous pastime at the best of times. I have always had my mainsails built so that the battens are horizontal to the boom, so that they remain in the sail as it is rolled. It sound a little drastic, but in practise seems to do them little harm,

The main disadvantage with roller reefing, is that when heavily reefed the boom tends to droop at the aft end. Some yachtsmen recommend a sailbag is introduced to the leech area as it is rolled away, but this is a messy hit and miss system that has little to recommend it. Far better to have the sail recut and remove about nine inches of canvas at the leech end of the foot to raise the boom slightly. This will reduce the sail area a little but it will be well offset by the convenience of a properly set sail.

The chances are that the reefing gear will be used frequently on a long cruise so it will pay to carefully examine and service all the moving parts of the system prior to leaving. There is nothing more exhausting than fighting a reluctant reefing gear on a pitching foredeck in a rising wind – I know I have tried it!

Chapter 5

Keels – Plain or Fancy?

Keels and Rudders

The latest fashion for cruising yachts in underwater appendages is the wing keel in its various forms. Since the late Ben Lexcen patented the wing keel, then fitted one onto *Australia II* and went on to win the *America's Cup*, the wing keel has arrived in a big way.

As is the case with many things designed to win races, the wing keel has proved its worth in the less frantic world of cruising. In fact many cruising designs had early developments of the wing keel some thirty years ago but it took Lexcen to see its potential and apply a sophisticated version to a pure race boat.

Way back in 1967 Van de Stadt the Dutch designers fitted a form of wing to their 30ft Arpege, it took the form of a bulb at the bottom of a fin, but in principal it worked the same way as the simpler wing, without an end plate. An American architect Henry Scheel designed what is now called the Scheel keel. Basically it is a thickening of the keel at the lower end to keep the ballast as low down as possible and many hundreds of yachts have been fitted with his design.

Development has been rapid since the success of Lexcen's *Australia II*, and the latest designs sport end-plates in various forms. The main advantage for cruising yachts lies in the fact that draught can be reduced without suffering the penalty of losing stiffness. The wing is also more efficient in that vertical tip losses are cut down with less water slippage so reducing leeway.

There have been many derivations over the years, some good and some bad. Take a stroll around any marina or boat yard at laying-up time, and you will see many examples of both. Unfortunately there are some examples of wing keels that are poorly designed, with just a crude end-plate applied to a stubby keel. In such cases the boat will suffer the disadvantages of extra drag and poor boat speed. If the wing is oversized it might even result in a tendency for the boat to broach in heavy going, but fortunately this is rare.

In general I believe a correctly designed wing keel on a cruising boat is an excellent option. It enables a yacht to be designed with a shallower draught, but to retain or indeed increase its stability. There is also sound evidence to suggest that some wings actually enhance a boats ability in rough seas and help its sail carrying capacity. The fitting of a ballasted end-plate, will increase a boats draught as it heels and tip losses are reduced. As these losses occur towards the back end of the keel's bottom, end-plates tend to be positioned here. In some cases they are delta shaped. In some comparative tests on wing and conventional fixed keels it has been shown that in light conditions the fin can be marginally faster, but once the going gets rough the wing simply romps away – which is nothing but good news for people cruising on the restless oceans.

A further development has been the tandem keel, designed by Warwick Collins.

This has twin foils and an end-plate. It has been designed to bridge the gap between competitive performance and reduced draught. Because the two vertical surfaces of the tandem keel extend the overall keel area fore and aft, directional stability is increased. The fact that the tandem keel is more stall resistant (up to 30° of leeway, as opposed to about 12° of leeway in most single keels) means that the improvement in directional stability and control in heavy conditions is marked,

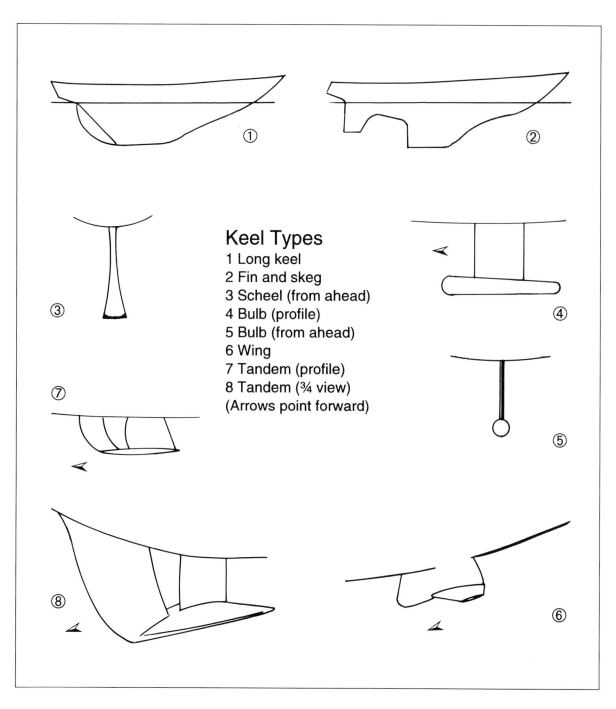

Keel Types
1 Long keel
2 Fin and skeg
3 Scheel (from ahead)
4 Bulb (profile)
5 Bulb (from ahead)
6 Wing
7 Tandem (profile)
8 Tandem (¾ view)
(Arrows point forward)

There are many different keel types that have been developed over the years. Some of the most popular used on cruising boats are shown above. All have their advantages and disadvantages, but for limiting draught without paying the penalty of reduced stiffness the wing, Scheel and tandem keels have much to recommend them.

Ocean Leopard, the 80ft Ocean class cruiser photographed just before her record breaking Round the Island Race.

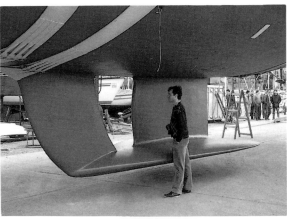

Weighing in at 16 tonnes, with wings 10ft wide, *Ocean Leopard's* tandem keel dwarfs its designer Warwick Collins.

which in turn reduces the likelihood of broaching. It is suggested that some sixty percent of the total weight of the tandem keel is stored in the main end-plate, which results in a righting moment equivalent to a conventional fin of up to forty percent greater draught. I certainly can support some of these claims from personal experience. In particular the reduced pitch and roll in a seaway due to the delta shaped end-plate, together with the uncanny directional control and stiffness in heavy going. Truly this is a keel for the man who wants to make fast ocean passages in comfort, yet also hankers after cruising the shallow out of the way anchorages in such places as the Caribbean and Bahamas.

The traditional full keel with rudder attached to its after section has its advantages too. These forms were highly developed during the '40s, '50s and early '60s until the fin and skeg designs took over. The main advantage with full keels is their resistance to leeway. Also boats with these underbodies are usually medium or heavy displacement, with typically fine sections foreword resulting in steady motion in a

seaway and reduced pounding when making to windward.

The disadvantages perhaps are increased wetted surface, and on some racing designs, where the designers have attempted to keep the wetted surface ratio down by reducing the keel length, skittish behaviour downwind. I am also less happy with some of these boats where the rudder post is excessively raked, this can also produce difficult steering when the wind increases. The majority of full keel designs however, don't suffer from these disadvantages, they are in the main well mannered and steady.

For many years I enjoyed sailing full keel boats, both racing and cruising, but having owned and crewed on many well designed fin and skeg designs, now prefer this type for a number of reasons. My principal concern when choosing a boat for fast cruising is balance, and it seems that modern well designed fin and skeg boats with sensible lateral area to their keels can be designed to balance more easily. They are generally faster in most conditions, and with a good sized rudder positioned well aft, they are easy to control. Of course there are

A modern underwater design with short lateral length of keel and spade rudder. The lack of depth at the forefoot will make her difficult to steer at slow speed in crosswind conditions when berthing.

A heavy displacement design with almost a full keel. I would expect this boat to have good directional stability and easy steering, but to be a little uninspired to windward.

exceptions and some of the more extreme racing designs that have cut the wetted surface to the minimum and fitted very high aspect ratio skinny keels, can be a handful if driven hard. For most of us however, the modern underwater sections of good cruising designs where moderation rules, are now perfectly adequate for long distance cruising. If married to efficient rigs the result is easily handled fast yachts that are fun to sail.

What is an important consideration to the blue-water cruising yachtsman, is the relative stability curve of his particular yacht. As a rough guide the deep, narrow, heavy displacement boats typical of '50s and '60s had a high stability factor. Their initial hull form stability was relatively low but as soon

as they heeled their low slung ballast would come into effect and the boats would stiffen up considerably.

With the influence of the IOR racing measurement in the early '80s, boats started to change shape dramatically. Light, beamy, shallow hulls became the vogue. Initial form stability, from their considerable beam was good, but as the boats began to heel it became necessary to get the crew weight on to the windward rail, to maximize the righting moment. Acceptable perhaps for racing round the coast, but dubious practise for long passages offshore.

When choosing a boat for blue-water cruising, it is best in my opinion to look carefully at the design parentage of each particular hull. A good type of boat is a medium displacement yacht with high positive stability for safety from capsize in extreme conditions. As a general thought the smaller the yacht, then the heavier the displacement (within sensible parameters) might be. She should also have a good length of keel for directional control. The larger the boat the more difficult it is to capsize, so lighter displacement, and more racier underwater sections can be considered. However moderation should always be the watchword.

There are dozens of well designed production yachts being built that are suitable for long distance cruising. It's a matter of personal preference which one to choose.

Rudders

On the question of rudder design I must admit to being old fashioned in my preferences. I want my rudder to be large enough for good control in all conditions, and bullet proof in construction.

This means that it must be securely attached to the boat, and in the case of a rudder that is separated from the keel, it should be protected by a full length skeg. I strongly disagree with some modern thinking where the rudder is designed with no skeg at all. This in my view is not a seamanlike design for an offshore yacht. The argument goes, that with modern materials, light displacement, sound construction and strong rudder stocks built deeply into the rudder, a skeg is unnecessary and disruptive to water flow, so slowing the boat slightly. It is also claimed that a skegless rudder can, if of the balanced or semi-balanced type give better, faster control downwind. This maybe so, but to my mind it's a crazy argument for a cruising boat. It may have justification on pure racing craft, with a large experienced crew of youngsters on board, but not on the average offshore cruiser which may be sailing shorthanded; here the integrity of the rudder is a major priority. One of my nightmares, along with being holed or dismasted, is to loose all steering through rudder failure. The oceans of the world are littered with all kinds of floating junk, from empty containers to large tree-trunks, just waiting for a yacht to run into them. If a boat is unfortunate enough to hit an object such as this, and there are regular reports of this happening each year, I want my rudder to be as strong and well

4 very different rudder designs

The rudders for offshore yachts should be as strong as possible. Separated rudders are vulnerable at sea, so the best designs are beefed-up to take the potential extra loadings.

Bad This type of rudder has been designed for speed and for this purpose is no doubt excellent, but for extended offshore work it is potentially lethal. It may be strong, but in collision with a heavy object, could snap off at the junction with the hull, or almost as seriously, jam over, so crippling the steering of the yacht.

Better The small skeg on this design gives some extra support to the rudder, but will not help it much if it is in a collision. In fact this particular yacht has just had some repair work done to the root of the fairing, caused by an accident.

Bombproof A truly massive rudder assembly on this motor sailer. The skeg is an integral part of the hull design and helps the tracking ability of the yacht. The twin engine installation , and the resultant wash onto the huge rudder, should give her outstanding manoeuvrability at close quarters.

Best A full length skeg such as this gives full protection to the rudder and enables it to be supported at both ends, therefore the whole assembly is very strong. It may reduce the speed of the yacht a trifle, but when offshore the difference would not be noticeable.

protected as possible to resist the impact. If the cost is a minute reduction in boat speed, then it is a price I am very happy to pay.

Should the worst happen and the boat loose its steering through rudder damage or cable failure, there are some measures that can be taken to get the boat back under control, although the thought of struggling to achieve some sort of emergency steering, possibly under difficult conditions in a seaway, does fill me with foreboding.

If the boat is tiller steered, then possibly the problem will be a broken tiller, so an extra tiller on board will solve this. If wheel steered, the cables could break (very unusual) or some failure in the quadrant mechanism on the top of the rudder stock may occur. An emergency tiller should always be carried (and tested for fit and efficiency well before setting out). We have drilled a large hole in the top quarter of our rudder and attached a short loop of rope through it. This will enable us in an emergency to attach two lines from the quarters, led to winches via turning blocks, which hopefully will get the boat back under control by us pulling on the lines to turn the rudder. We have used rope as a loop rather than a large shackle. (It would need to be fitted too close to the top edge of the rudder.) Wire should be avoided, as this would

quickly saw its way through the rudder. Another idea we have considered is to utilize the spinnaker boom as a steering oar over the transom. For this to work would mean constructing a large blade on the end to act as a paddle, and lashing up a suitable rope 'rowlock' on the pushpit. To this end I have pre-drilled two large locker lids from the aft cabins to fit the spinnaker boom end, where they can be securely lashed with wire. To make this oar effective, it would have to be weighted at the 'wet' end, and for this, we would use the kedge anchor, suitably padded and lashed. To ensure this contraption would work reasonably efficiently, the boat would have to be balanced carefully with the sails, but in our case with a ketch rig this should not be too difficult – however, I hope it will never be put to that test!

There is a piece of equipment available that it is claimed is capable of steering a yacht by trailing a drogue astern on a long warp. It is called the Attenborough Sea Drogue™ and being constructed of stainless steel is strong enough to act as a sea anchor as well. I have not seen it in action but it certainly is well made and for yachtsmen who favour sea anchors it could be a worthwhile piece of equipment to have in the ship's locker.

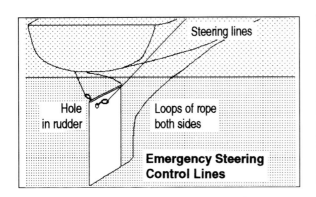

Hole in rudder

Loops of rope both sides

Emergency Steering Control Lines

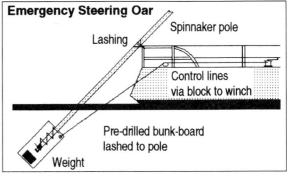

Emergency Steering Oar

Spinnaker pole

Lashing

Control lines via block to winch

Pre-drilled bunk-board lashed to pole

Weight

Chapter 6
Seaworthiness

Heavy Weather Management. Anchors and Anchoring. Ropes.

Seaworthiness is a function of good design, that has taken into consideration such things as stability, sail to displacement ratios, freeboard, buoyancy and so on.

The great majority of production cruising yachts are perfectly capable of blue-water passages; provided they are equipped and modified to the common sense standards required for such an undertaking. The size equation applies largely to the degree of comfort, or perhaps I should say discomfort, one is prepared to accept; for in my view there is no doubt that the relationship of size to comfort is very close. The larger the boat, within reason, the more stable and comfortable she will be offshore, and come to that, at anchor as well.

Ability at sea is another matter, but here also size has its advantages. Turning to windward in strong conditions is a masochistic exercise and no one in their right mind can actually enjoy the experience for any length of time; but as in most things in life it's all comparative. Take a small yacht of say 25ft LOA with a wind of Force 5 on the nose. She will be well reefed and sailing on her ear; to her and her crew these conditions are close to a gale. Whereas the same wind in an able 50ft LOA yacht would represent just a good wholesome breeze.

The same parameters apply to the seas one faces offshore. To get a really small yacht to windward in a strong blow can be a thankless wet task, where progress will be painfully slow. She will lack drive due to her sails having to be well reefed. With a large, more powerful yacht, she will have to reef less and this, coupled with her inherent stiffness and stability will enable her to maintain her speed in relative comfort. I can hear some racing yachtsmen disagreeing

with this opinion, and siting their own experiences in driving their small yachts to windward in such a such race in winds of Force 7 or more. I quite agree with them, it's quite possible but its horribly uncomfortable and makes no sense offshore when one is faced with perhaps several days of such conditions. It is in such conditions that size becomes important, for I equate ease of motion and stability, directly with comfort, and the ability of the crew to be fed and rested adequately – and this ultimately effects the seaworthiness of the yacht, for exhausted skippers and crews make more mistakes than well rested ones.

Much is made by yacht salesmen and owners of the relative close windedness of their particular boats. For a blue-water sailor this matters very little provided of course, the boat in question is a well designed yacht with a reasonable ability to windward. The few degrees closer a racer can point to windward is really of little consequence on the ocean. Quite often we will bear away to free our course, to enable us to keep boat speed up. This also helps to give us a comfortable ride, and more importantly, enable the cook to keep his feet and serve up more appetising meals. The extra miles sailed don't add up to much either, when perhaps the next day the wind will free and let us sail the rhumb line again. But of course this is not a matter of size or seaworthiness, but really just common sense seamanship.

Heavy Weather Management

Every long distance yachtsman must face the fact that he will face severe weather sooner or later. Strong gales at sea are part of the cruising scene, but if care is taken to select

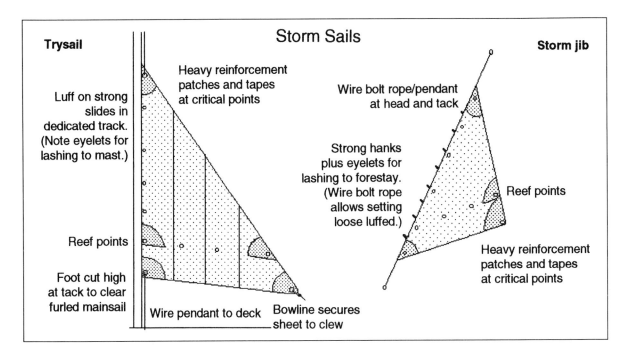

Storm Sails

Trysail

Luff on strong slides in dedicated track. (Note eyelets for lashing to mast.)

Heavy reinforcement patches and tapes at critical points

Reef points

Foot cut high at tack to clear furled mainsail

Wire pendant to deck

Bowline secures sheet to clew

Storm jib

Wire bolt rope/pendant at head and tack

Strong hanks plus eyelets for lashing to forestay. (Wire bolt rope allows setting loose luffed.)

Reef points

Heavy reinforcement patches and tapes at critical points

Storm sails should be included in every long distance cruising yacht's sail locker.
If the trysail is of a largish size, with reef points, it can be used as an effective hard weather mainsail.
About 30% of the normal mainsail area is considered right.

routes and departure times correctly they can, with a little luck be largely avoided.

Gales at sea can be very frightening, but in a well found yacht properly managed, rarely dangerous. It's the incessant noise of the screaming wind, the frightful motion of the boat, the build up of exhaustion from lack of sleep and perhaps seasickness, that all contribute to a feeling of lassitude which can be the real problem. If a gale is a long one, the mind seems to go into neutral, where thinking logically is difficult and to work out the simplest navigational problem becomes almost an impossibility.

I have discussed this with a number of experienced yachtsmen, and they have all, without exception, felt the same when conditions really deteriorate. So what is to be done?

The first requirement of course is the boat itself. If you are convinced that you are in

for a hard blow, reduce sail sooner rather than later. Make sure all gear above and below decks is securely lashed down, and the boom well secured in a gallows or tied down to the deck. Haul out the storm sails and have them clipped on ready for hoisting, but ensure they are securely tied so that any loose water on deck will not wash them overboard. Give all the crew a hot nourishing meal, not a snack, for it could be a long time before any more serious cooking might be possible. Issue the whole crew with seasickness pills, even the hardy sailors, for very often sickness is brought on, not by motion, but apprehension. Check the bilges and pumps, and bring the navigation right up to date, in order that the precise position of the yacht is known. Enter all the relevant data in the log. Make sure all harnesses and life-jackets are available for immediate use and ensure the deck watch wears them all the time when on deck or when below. Then encourage the off watch crew to rest in their bunks, if not asleep then reading or just relaxing in anyway they think fit. The energy of the crew is a vital factor and later their reserves of strength might be crucial.

Once the ship is well snugged down there is a temptation just to sit and wait, but don't be tempted, keep active and sail the yacht as normal. Try to gain as much sea room as possible between you and the approaching storm. Sea room is the first fundamental of heavy weather management, for to face a gale well offshore with plenty of sea room is a totally different proposition to one close to land with all its extra problems and dangers.

I well remember some years ago putting my yacht and crew at risk through an error of judgement. It was late in the season and we were storm bound in Weymouth after a West Country cruise. The weather was so bad we decided to leave the boat and return home, to be collected the next weekend. That year the equinoctial gales were well into their stride and it was another four weeks before a weather window was forecast, that coincided with a weekend. By then it was early November and I was desperate to get her back to her home port. The forecast was unsettled but with winds around Force 6 from the North-West. I calculated it would be a quick dash back before the next front was due. Unfortunately, it was a period of exceptionally high spring tides. I planned our run to arrive at the Needles, (a particularly nasty place in strong wind over tide conditions) at just after low water, then carry the flood all the way to the Chichester bar, hoping to arrive around high water.

In reality the wind was nearer Force 8 than 6, but I decided foolishly, as all the arrangements had been made, to carry on and sail. We made very good time across Weymouth Bay, but I soon realised the weather was deteriorating further, and a call to Portland Coastguard confirmed Force 9 in the gusts with further deterioration forecast for later.

It was then I made my second error based not on experience but of optimism. I knew this coast well and realised that we could not turn back, or heave-to, as we were too close to land. St Alban's race was between us and the relative safety of Poole Harbour, so I decided to try to find an inshore passage; which should, I hoped, give us some respite from the growing seas. The conditions of course were impossible for the inshore passage to exist, with a very strong ebbing tide of perhaps four to five knots against strong gale force winds. The seas were a mass of white tumbling crests, due to the shallow water.

We decided it would be sensible to call the coastguard, who quickly confirmed the lack of an inshore passage and warned us of the very dangerous conditions that were prevailing in the race at that time. He strongly advised us to get at least eight miles offshore. This we tried to do, but I had left it too late, and we were drawn into the race, which extended perhaps two or three miles further west than normal, due to the exceptional conditions.

We had to make the best of it and get through as well as we could. The seas were quite amazing at this time, not particularly big, twenty feet or so, but with almost vertical faces, they were very unstable and crashed down all around us in a smoulder of foam. Another interesting feature was that they were completely irregular in direction, coming at us from several angles, all at the same time, which made steering particularly difficult.

We were down to storm jib, but soon even this tiny sail had to come off, for as *Zola* dipped into a hollow, she would lose most of the wind and steerage way, swing round broadside-on, then the sail would flog, like a dog shaking a rat, before the next crest would hit us broadside-on and carry us in a semi broached state, until she was brought under control again. It was around this period we started to take heavy water on board and were pooped several times in quick succession. I realised trying to run under bare pole was useless. So I did something I have never done before, I started the engine to maintain steerage way, and it worked! We were able to keep the stern more or less facing the seas and although Simon, who was steering, disappeared completely under water from the breaking seas from time to time, we were at least under some semblance of control. Unfortunately due to our slow speed through the water, and the strong ebbing tide we were locked in the same position for several hours, but eventually the tide turned and we crawled painfully from the grip of those awful seas.

From this experience another few lessons were learnt. Never put to sea, whatever the pressure, if it is unseamanlike to do so. Never hope conditions will improve, make decisions based on common sense and experience and most importantly of all, keep as far away from land as possible when bad weather is forecast. Although it can be a hard decision to make, if faced with an offshore gale; put to sea rather than run for harbour, you will be much safer and less likely to put your boat or crew at risk.

Most gales are relatively short in duration. In summer perhaps 12-18 hours, for the fronts usually are fast moving and they clear quite quickly, often to set seasonal patterns. If you are caught in a storm at sea with limited sea room what is the best tactic to adopt?

The main priority is to keep moving away from the area as quickly as possible. This can be achieved by sailing under storm canvas, or motor sailing under deep reefed main or trysail. If the wind is directly ahead and too strong to keep sailing then heave-to, on the offshore tack. Boats heave-to in a different manner depending on type. The deep draughted full keel yacht will usually lie very quietly with storm jib aback and helm lashed down, fore-reaching at about a knot or so. Other boats will range about a great deal and perhaps need a touch of mainsail up; whilst others like a boat I once owned, would heave-to under deep reefed mizzen and storm jib. It is important to find the combination that suits your particular boat, well before facing heavy weather offshore. The characteristics of heaving-to can be judged just as easily in a Force 5 or 6 one weekend before embarking on a trip, and the lessons learnt used in earnest if required at a later date.

Stage Two, Lying Ahull.

If the wind continues to increase and starts to overpress the yacht the next stage might be to lie ahull. Simply take all sail off the yacht and let her take up her natural position in the water. This will greatly depend on her underwater shape and her resistance to leeway. With some light displacement boats the motion can be violent but all should be well if the rudder is lashed slightly a-lee, but again this might call for some experimentation. When a yacht lies ahull, it appears to leave a calm slick to windward which can help the motion a little. The danger point in this procedure occurs when the seas build up and start to break. Believe me, this is soon recognised as the boat will suddenly drop down the seas as in a lift, then with a rush and a roar the breaking sea will crash aboard and push the boat bodily to leeward. This is the danger sign that lying ahull has now become dangerous.

Heavy water on board can break gear and fittings very easily, for contrary to popular belief water is a very heavy substance. One cubic metre of sea-water equals about one ton of mass, if you calculate how much comes aboard from only a relatively moderate breaking sea it's obvious that boats can be severely damaged in these circumstances.

The other danger with lying ahull in large breaking seas, is that the boat is in a vulnerable position for being rolled over, as the waves tumble forward with a wall of water cascading down their face. The boat will trip over its keel lying in the deeper more static water and roll.

Stage 3 Running Before a Storm.

When conditions are so bad that lying ahull is no longer possible, then the boat must be run off down wind. If there is not a great deal of sea room to play with, slow the boat down as much as possible by trailing warps, anchors, tyres or anything that will create enough drag to keep the boat sailing slowly, but under control. (See Addendum p.164).

If sea room is not a problem then the yacht can be sailed at a speed to maintain comfort and safety, either under bare poles, or with a storm jib or trysail hauled amidships. The trick is to keep the seas from breaking aboard, by moving ahead at sufficient speed, and taking them fine on the quarter. We have run like this perfectly comfortably in heavy conditions with dry decks and little strain on the yacht. If the speed of the yacht starts to interfere with the seas aft, by making them break, then slow the boat up by towing a long nylon warp in a bight from each quarter. This warp should ideally be about three to four hundred feet long to keep it buried in the seas astern, but not all yachts have such a long length of line on board. If this is the case, use the longest you have, or perhaps use two and weigh them at their apex with an anchor or other heavy item to keep them under the water. It is quite surprising how this slows the boat, yet enables the crew to maintain steerage way.

It is important to find out what an individual boat requires in terms of gear and technique to bring it through a storm safely. Some experienced seamen prefer always to run fast, with the seas on the quarter, others like to slow right down taking the crests on their sterns, whilst others (now relatively few) swear by holding their boats head to sea with a sea anchor. The problem with this method is that it imposes heavy strains on the rudder and its fixings as the yacht makes sternway.

For most of us, these 'ultimate' arguments are thankfully, really academic, for we should not have to face such conditions if we cruise our boats to the sun during the well documented 'safe' seasons. However it is prudent to think about, (and test our boats in less arduous conditions) on the methods that we will adopt, if we are unlucky enough to get caught out in really severe weather.

Anchors and Anchoring

Talk to any ten skippers and you will get ten differing opinions as to the best anchor and the way it should be used.

The fact is that we all have our favourites, which are based on previous experiences, some good and some bad. Some of us tend to develop an anchoring technique that suits our own particular ground tackle and yacht.

In some forty years or so of sailing in northern European waters I don't suppose I have dragged seriously more than a dozen times, perhaps twenty at the outside, and most of these were when I probably had just dropped the hook quickly for lunch.

I therefore consider myself a reasonably competent skipper who understands how to anchor a vessel safely. When I started cruising in the Mediterranean we dragged on the first attempt and subsequently drag one in every three or four times we try to anchor, even now. My self esteem took a hammering and confidence dropped until I rationalised the problem. Essentially in northern European waters two factors work for you. One, the tide and possibly the wind, and secondly the type of bottom most likely to be found – usually clay or mud which gives excellent holding. An advantage also is that depths in general are on the shallow side so plenty of scope can be given.

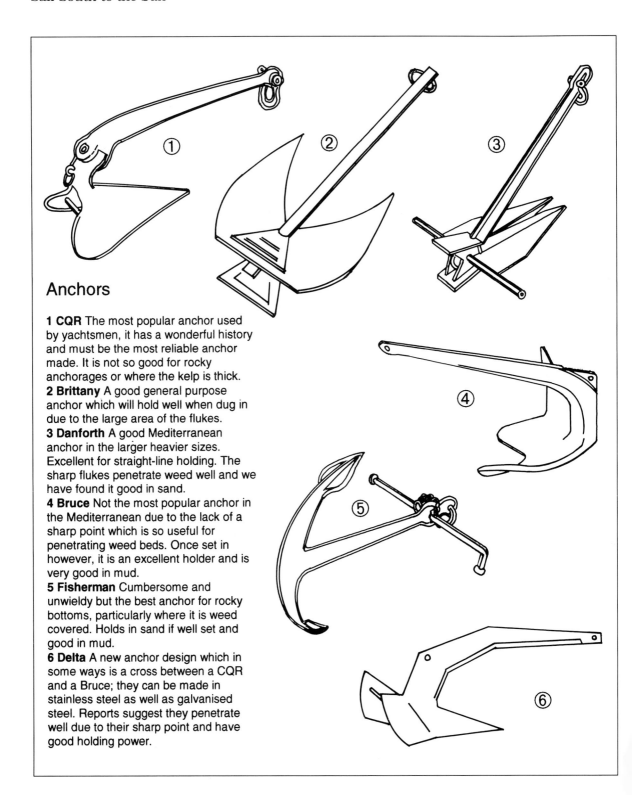

Anchors

1 CQR The most popular anchor used by yachtsmen, it has a wonderful history and must be the most reliable anchor made. It is not so good for rocky anchorages or where the kelp is thick.

2 Brittany A good general purpose anchor which will hold well when dug in due to the large area of the flukes.

3 Danforth A good Mediterranean anchor in the larger heavier sizes. Excellent for straight-line holding. The sharp flukes penetrate weed well and we have found it good in sand.

4 Bruce Not the most popular anchor in the Mediterranean due to the lack of a sharp point which is so useful for penetrating weed beds. Once set in however, it is an excellent holder and is very good in mud.

5 Fisherman Cumbersome and unwieldy but the best anchor for rocky bottoms, particularly where it is weed covered. Holds in sand if well set and good in mud.

6 Delta A new anchor design which in some ways is a cross between a CQR and a Bruce; they can be made in stainless steel as well as galvanised steel. Reports suggest they penetrate well due to their sharp point and have good holding power.

In the Mediterranean the opposite applies. Depths are considerable, the bottom is usually soft sand or weed covered rock and there are no real tides to assist anchoring. Winds are also somewhat of a problem at times as they tend to move around, sometimes by up to 180° in one shift. On several occasions we have anchored well clear of obstruction in a lovely cala or cove yet within a couple of hours the wind has veered 180° to put us close onto a rocky lee shore. We have learned now to always allow for a 360° swing if anchoring for the night, and if this is impossible due to restricted swinging room, we use a Bahamian moor, or in very light weather lay out a second anchor to hold us by the bow and stern.

One of the main irritations with Mediterranean anchorages is the eel grass weed that covers the bottom in many places. These rafts of weed can be several feet thick and at times it defies an anchor to penetrate it, resulting in the boat just being held by the weed mat, which is most dangerous. Other bottoms are rock with just a thin layer of weed, which is almost as difficult to cope with. The anchor slithers along the bottom rock until it finds a crevasse to wedge itself into; perhaps just the extreme point is holding the boat which again is less than satisfactory. Sand, although our usual choice, can also be treacherous on occasions, for in some areas it is very light and powdery so the anchors have very little to dig into – so they drag, perhaps picking up a clump of weed as they trawl along the bottom. The only thing to do then is to raise the anchor and try again. It is not unusual to see boats try two or three times before they are satisfactory anchored.

Of course it's not all bad, it's just that the conditions are different in the Mediterranean and the prudent skipper should adjust his technique to cope with all the conditions he is likely to face.

The anchoring routine we use has been learnt the hard way, and on most occasions we now manage to get the anchor to hold first or second time.

Firstly, we motor slowly around the selected anchorage with a crew member on the bows, checking out the bottom. (The water is usually so clear that the bottom can be seen up to around thirty feet or more.) Once the ideal spot has been selected (usually over sand) we make a mental note of the depth then drop the hook as close to the ideal position as possible, with the yacht making a slow stern board. Once the anchor has hit bottom we slowly pay out more cable being careful not to foul the anchor. I like to have about a three to one depth ratio of cable to depth, ie 30ft:10ft depth at this time. I then gun the engine a little to dig the anchor in, and if all is well continue astern until the ratio is five or even six to one depending on conditions and space available. At this point I put the engine full astern for a long burst, to put maximum strain on the anchor. The cable should then stretch out straight from the bows if the anchor is holding, then relax back to a steep curve just under the bow once the power is turned off. This might sound a little drastic but if the anchor will not hold the boat with the engine full astern, it's a certainty it will drag if the wind picks up.

Even when I am happy that the boat is securely anchored, I always send out a swimmer with fins and mask to check that the anchor is well bedded and not fouled in any way. In the clear, warm waters of the Mediterranean this is never a chore and many times a refreshing pleasure after a hot session in the sunshine.

We have developed a straightforward system of hand signals to convey to the helmsman just what is happening on the bow or below the yacht, and it is most gratifying on occasions, to complete what is perhaps a tricky anchoring manoeuvre without a word being spoken – or should I say shouted!

One problem that is relatively common, is

It is a good plan to position someone on the foredeck when approaching shallow or coral anchorages.

A well designed foredeck arrangement with strong anchor bitts and cleats.
The double stainless steel bow chock has a drop-nosed pin for locking the anchor in place when the yacht is at sea.

getting the anchor fouled by rock or some other underwater obstruction. This is usually more embarrassing than terminal, but it can prove difficult to solve at times. The correct method of anchoring in suspect areas is to always use a trip line secured to the anchor crown with a small buoy at the 'top end'. This also has its difficulties; as sometimes the depths involved are considerable, which would call upon a very long trip line to be used. In general I don't use the system very often, as the line seems to get tangled up with the chain or round the keel or rudder more often than not. It is also a danger to other yachts, wind surfers and the like. They seem to love sailing close to anchored boats for some reason. I am also aware that if a trip line and buoy are out overnight, an unsuspecting yacht can easily get tangled up with it, so tripping our anchor and probably disabling his propeller with particularly unpleasant consequences.

If you are unfortunate enough to get the anchor fouled, bring the cable in until it is very short up to the stem head, then get the crew to move aft. This sometimes is enough to lift the anchor clear. If not, slack off a fathom or so then motor *slowly* round in a circle keeping the cable as tight as possible. If it still won't move, try motoring *into* the centre point of the circle so that the boat is brought up with a jerk as the cable becomes taunt; this should shock the anchor clear of the seabed, but be careful not to use too much power or the cable could break, or indeed the samson post or mooring cleat be damaged. If all these efforts are to no avail, then the anchor should be buoyed, and the cable slipped, to be recovered another day with the help of a diver. If the anchor is a good one it will be well worth the trouble and cost of rescue, as replacements in the Mediterranean are difficult to come by, and expensive.

Storm Weather Anchoring

It's conceivable that at times, every cruising boat will have to face a day or two anchored in storm conditions somewhere. When faced with this, it's little use wishing for heavier or more adequate gear. The wise skipper makes sure this has been a priority when fitting out the yacht for the cruise in the first place. For Mediterranean or Caribbean cruising in the 'summer' seasons, fewer anchors and perhaps slightly lighter gear can be used. If however plans are in hand to cruise extensively and live on board for long periods, the ground tackle should be heavy and comprehensive.

Our compliment of anchors is, if anything, rather on the light side. On board is a 45lb CQR bower with 35 fathoms of calibrated chain. One 35lb Danforth on 3 fathoms of chain spliced to 200ft of braided nylon, plus the heaviest Fisherman anchor one man can comfortably handle, which weighs in at about 60lbs. This is also shackled on to 3 fathoms of chain and 200ft of nylon warp. I made a mistake when fitting out by not replacing the CQR with a heavier model, but propose to amend this when the opportunity presents itself. To back up these anchors for an emergency, we have a spare anchor warp, of 250ft heavy nylon which has in fact been used on one occasion this past year. Handling heavy equipment is always a problem and some form of power or hand windlass is necessary on all boats from about thirty five feet upwards if back strain is to be avoided.

When anchoring with chain in severe conditions try to run out the maximum amount possible, the catenary or curve the chain takes, will help to avoid snatching in the stronger gusts. If the weather is really bad or, because of restrictions on room you are lying to a short scope, lower a heavy weight halfway down the chain on a warp. This will act as a dampener and keep the shock loads to a minimum as the chain straightens out. Another excellent way to reduce snubbing is to use a heavy claw that hooks across the chain links, this is in turn shackled to thirty feet of heavy nylon line, which is made fast to a cleat or samson post. The nylon being stretchy, acts as a shock -absorber. It is particularly useful on boats that have windlasses, as the strain is taken by the nylon line and removed from the windlass gipsy. These hooks are now being manufactured again by Sinclair Everitt and are available in a range of sizes.

Laying out two or more anchors when facing difficult conditions is straightforward, but the technique needs thinking through well in advance. I prefer to use chain for my bower anchor, due to its strength, weight and self stowing attributes. When faced with using two anchors we use the following procedure.

After selecting the place to anchor we drop the bower, and get it well dug in with the maximum scope run out. We then row out the kedge at about 30° to the main anchor, then drop it, using the maximum scope possible. Once the kedge is well dug in, it is adjusted on board to just take the main strain at full stretch, before the bower and chain take up, as a back up role. If the pressure is too much and the kedge starts to drag, reverse the rolls of the anchors and use the bower as the primary anchor. It is important to carefully parcel the nylon rope where it enters the fairleads or chafe will quickly get to work, with disastrous results.

If the weather is too rough for the dinghy to be used, then the yacht will have to be manoeuvred to drop the two anchors. This is perfectly feasible. To successfully carry out this manoeuvre ensure that the anchors and

A wide variety of anchorages is available to the cruising yachtsman in the Mediterranean, from tiny calas –

– to large open harbours

warps are carefully flaked out on deck for free running, before starting.

First motor to the chosen spot and drop the first anchor, then fall back with the wind until the anchor is well set, then the required amount of chain or warp is run out. Once you are confident you are holding, motor forward again at an angle of about 45° to the main anchor and drop the second one in line with the first. Again drop back with the wind, playing out the warp as you go, until the first line is taut, then make fast and you should be lying equally between the two anchors. They can then be adjusted at leisure to equalize the strain.

Another method of backing up a bower, is to anchor in tandem using two anchors one behind the other. The linking chain should be about three fathoms long and fastened to the anchor stock at its base. To lay, first drop the forward anchor, then once it has reached the bottom follow it with the second, and slowly go astern until they are set. It is a complex manoeuvre that needs practise, but although I have never had much success myself, it comes well recommended by other experienced and perhaps more skilled yachtsmen than myself.

There will be times when anchoring in some parts of the Mediterranean when you will be faced with a bottom that slopes steeply off and it will be difficult to anchor close to the shore. In these cases the best course is to take a long line ashore from the bow then anchor with a kedge from the stern.

Scopes

In general, a guide to anchoring scopes in light weather is to allow three times depth for chain and five times depth for nylon line (with two to three fathoms of chain at the anchor end).

In rough weather increase the ratio to five or six times depth for chain and eight or nine times depth for nylon line.

These are approximate guides, for different conditions, alter the scope as required. With chain, the boat will move about less in the wind due to the damping effect of the heavy chain, where as nylon rope being that much lighter will not have this advantage. The big safeguard with laying out a long scope, is that the pull on the anchor will be more horizontal in its effect, so keeping it well dug in and enabling perhaps smaller anchors to be used. With shorter scopes the cable will be less horizontal so the anchor is more likely to be tripped.

Many anchorages facing the open sea, both in the Caribbean and Mediterranean, can suffer from a persistent swell which although not necessarily severe can make them uncomfortable.

A ketch rig has the advantage here over the sloop in that it can hoist its mizzen which will bring the yacht into the wind and help dampen out the swell a little. Sometimes a kedge anchor laid out from the stern in line with the prevailing swell can work wonders and is often worth the effort and trouble.

Some thoughts on Rope

Perhaps it might be of help to clear some of the confusion that surrounds the ropes we use on our yachts. There is a wide variety of materials, constructions and brands available, and some are designed to be used for specialized purposes.

Just to go back a little in time, after the last World War a revolution took place that totally transformed the manufacture of ropes, it was based on replacing natural fibres used by the industry for centuries, with a new material, synthetic fibre. The main advantages with this new material were much longer life, considerable gain in strength and a resistance to rot; even when stored under damp conditions.

What a gift this was for us yachtsmen, no longer did our sheets and running rigging stretch, we could rely on ropes not to break through rot and their life was almost infinite provided they were not allowed to chafe. The new ropes were also more pleasant to handle and because they did not absorb much water, much lighter too. Technology has moved along a pace, and now we have a whole measure of additional advantages, such as colour coding, self colouring, floating ropes, non- stretch ropes, high-stretch ropes, soft handling

ropes, non-kinking ropes and so on. All this is just wonderful, but just a mite confusing at times, when we are faced with row upon row of reels to choose from at our local chandler, so a little classification might not go amiss.

Rope Materials

Basically there are four materials used in synthetic rope construction; Polyester, Polypropylene, Nylon, Kevlar™ and four also in traditional natural ropes; Hemp, Manila, Sisal and Coir. All have their advantages and disadvantages. Modern yachts rarely use the natural fibre ropes these days. Natural fibre ropes are usually the province of the purist who wishes to maintain his yacht in the traditional manner.

The Synthetic Ropes

There are a number trade names or brands which usually have their own colour coding style to differentiate them from their competitors, but essentially they are all similar in performance. In addition to colour, there is, on some materials a choice of matt or glossy finish depending on what the rope is to be used for.

Polyester

This is the most common material used in synthetic rope construction, and it is very strong and versatile. In the main it is used for sheets and halyards. It can be 'prestretched' in its three strand form during manufacture, so is ideal for halyards which require the minimum of elasticity. It comes in braided, plaited and three strand forms.

Polypropylene

A cheaper rope with less strength than polyester or nylon. It is also less comfortable to handle. I have a coil on board because of its ability to float, which can be useful at times. I have seen it used as an anchor warp but feel this is dangerous, for it can become tangled in the propellers of unsuspecting boat owners who stray too close. It can also be used for mooring lines in its larger sizes. It is vulnerable to UV degradation, so check out that it has a UV inhibitor added during manufacture.

Nylon

Wonderfully strong with the important attribute of being able to stretch to approximately forty per cent before reaching breaking point. Excellent for anchor warps and mooring lines but not for halyards or any other job that needs non-elasticity. When used for mooring or anchor rodes, care must be taken to avoid chafe in fairleads etc.

Kevlar

Very strong, very light and very expensive. The racing man's rope, where expense is no object. Its light weight to strength ratio means it can be used in place of wire for some applications. It can be spliced and only loses ten per cent of its strength. Being a stiff rope it kinks easily and doesn't like being taken round sharp corners or bends, so extra large sheaves should be used.

Natural Fibre Ropes
Manila

One of my favourite ropes from many years ago. It has good resistance to rotting when wet. Can be handled easily and is strong for its size. It wears well too.

Hemp

Unmistakable in its aroma. This is a lovely rope, though difficult to splice. Excellent wear and strength properties makes it a jack of all trades on traditional boats.

Sisal

Sometimes a difficult rope to work, weaker than manila and when damp tends to get rather stiff.

Coir

The old fashioned rope that is familiar when used for those decorative fenders that weigh a ton when wet! It's very prickly to work with, but when dry, is light and stretchy. It was used for mooring warps before nylon came on the scene. Still used for decorative rope work.

Rope Construction

All manner of synthetic rope construction is available now but to most of us, the four familiar ones are:

Braided, Plaited, Eight Plait, Three Strand.

Three Strand

There are a number of different 'lays' to accommodate the use the rope will be put to. Right hand or 'Hawser laid' coils in a clockwise direction – and left hand laid is also known as an S twist. Cable laid is constructed with three strands and is rather more pliable than two strand rope. When used on nylon in particular these lays are very easy to splice using hard eyes or on chain to rope applications. Can be hard on the hands.

Braided

A lovely construction than can also be laid as braid which means the rope has an additional braided core. Ideal for sheets due to its strength characteristics. It is a 'bulky' rope with a large surface area so it works well on winches. Splicing this rope is beyond my capabilities so I always use a professional, otherwise it is a great rope, that is a joy to handle.

Plaited

Multiplait is another name for this rope as its construction is eight separate strands plaited together. I use it as a general purpose warp as it is easy and soft to handle. It also coils easily and does not kink, perfect when used with nylon as an anchor or general purpose warp.

Eight Plait

This can also be purchased as sixteen plait. It makes a soft supple rope that is used in the main for halyards and sheets. I have used it for many years in polyester and have great confidence in it. The matt finish in particular is smooth to handle and as such is ideal for use as sheets, as it will resist 'slip' on the winches.

Even the best synthetics wear in time but to maintain maximum life always be aware that chafe is their greatest enemy. It's well worth the effort each season to turn halyards end for end, or if this is too difficult chop twelve inches or so off the shackle end so that the 'nip' in the sheave falls in a different place each year. With sheets it's simple to reverse the ends each season, but also ensure the rigging screw split pins are well taped to avoid catching the sheets. Anchor and mooring warps will quickly chafe through in the Mediterranean where there is an almost constant swell. Cut short lengths (two to three feet) of plastic hose and thread them over the warps (don't split them, as is sometimes recommended) where they come aboard and through fairleads. Drill a couple of small holes near to the top end of the plastic hose, and use a short length of marline to tightly bind the hoses to the lay of the warp. This will keep them in place and double the life of the warps.

Chapter 7

Getting Her Ready for Blue-water Cruising

Watertight Integrity, Seacocks, Cockpits & Companion-ways, Hatches & Portlights, Bilge Pumps, Working the Decks, Jackstays, Harnesses, Caring for the Sails.

Watertight Integrity

The very heart of a yacht's seaworthiness lies in her watertight integrity. Unfortunately yachts have to have a number of openings and holes in them to make them habitable and therein lies their weakness. It is not however, beyond the average owner to analyse these weaknesses, then set about making them as secure as reasonably possible.

The many attributes of a yacht's design that make her comfortable and pleasant to live in during a cruise in warm waters, such as big cockpit, lots of hatches, open-plan airy interiors etc, run contrary to the ideal layout for offshore cruising. So as always, a compromise has to be struck.

Assuming the boat is soundly constructed and in good condition, it should not be too difficult to modify her to make her suitable for cruising in blue-waters. Minor leaks and drips are quite usual offshore when a boat is working hard, and most boats suffer from these to some extent.

Hatches and skylights that are normally quite tight, can, in gale force conditions, when rain and spray are driven by the wind, leak quite badly, usually in the form of a light mist forced in between minute cracks between the seal and the surround.

These problems are relatively minor and can be dealt with during routine ship maintenance. The real potential for serious flooding lies in the major deck, hull and cockpit openings and these must be given serious consideration before leaving on a long open water cruise.

Seacocks

All seacocks should be checked and serviced before departure. If they are of the cone type, regrind them, then grease them well to ensure they can be shut off easily. Carefully examine the through-hull fixings. If there is sign of corrosion or electrolysis (white powder deposits) remove them for replacement.

Double clip with stainless steel jubilee clips all the hose connections. If in doubt about the security or quality of the hoses in any way, replace them with new. The easiest way to get a tight hose onto a bronze spigot is to immerse the hose end in very hot water for a minute or two until it is soft and pliable, then slide it on and tighten up the clips whilst the hose is still warm. If you still have difficulty, apply a little neat washing up liquid to the spigot to reduce friction. Don't forget the engine cooling water inlet and the cockpit drains, they should all have seacocks fitted. Once all the seacocks have been serviced, tie a tapered, soft wood plug to each, for use in an emergency.

Cockpits

Cockpits are vulnerable to flooding from breaking seas shipped over the stern. Centre cockpits are safer in this respect, but suffer the disadvantage of being closer to the bows, so in consequence are wetter when going to windward.

The main areas for attention in the cockpit are – lockers, companion-ways and mainhatches. If the boat has large opening

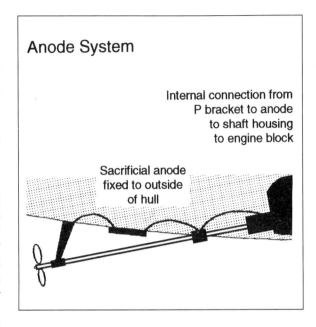

Anode System

Internal connection from
P bracket to anode
to shaft housing
to engine block

Sacrificial anode
fixed to outside
of hull

Due to the warmth and high salt content of the water, electrolysis can be a problem in the Mediterranean. It is therefore very important to ensure all external metal is well protected with a sacrificial anode system, this is especially so with aluminium and steel hulls.

If the engine has a sacrificial anode in the block, change it annually, particularly if the engine is raw water cooled.

lockers, firstly ensure that they have efficient watertight seals, with a drain runnel built in round the edge. They should also have strong hinges and a substantial method of locking them shut. On some production yachts these fittings are flimsy and could prove a potential danger. Look at yours with a critical eye and if at all doubtful, replace them with stout fittings that will stand the pressure of solid water. If a locker top is lost overboard due to its hinges or fittings giving way, the boat is opened up to the seas and potential flooding, for it is an opening that will prove extremely difficult to plug quickly and efficiently.

Companion-ways and Cockpits

Companion-ways are better if they have a bridge-deck to protect the interior from an influx of water. On most modern boats the weather boards are made with a slight tapered wedge shape to enable quick fitting. Ensure they will stay in place if the worst happens and the boat rolls over 180°, by fitting them with some method of retention from inside the saloon. On *Mishka* we have bolts to hold them in place, which seems to be effective. If the top board is solid, it is worthwhile installing a ventilator here.

There is not a great deal that can be done to a main hatch if it has been incorrectly designed, other than replace it. A good watertight hatch should have its own garage when slid forward, with high upstands and drainage holes.

Cockpit drains should be as large as practical. A minimum of two x two inch drains for a thirty foot boat. It's a sad fact nowadays that many modern yachts sail around with minuscule cockpit drains which are only of use to clear rain water. These are totally inadequate for offshore work. If flooded, the cockpit must be able to clear itself quickly, for if it doesn't, the water will pull the stern down and unbalance the boat. This in turn makes it more likely that the next following sea will break aboard even more heavily, perhaps washing the crew overboard or bursting below through the companion-way. With all this extra weight the boat will lie like a half submerged log and be very vulnerable to being rolled over.

Having experienced a situation very close to this myself some years ago, I am particularly anxious not to repeat the lesson. It's too late to agonize over the fact that the drains are inadequate when offshore, and facing heavy breaking seas; much better to ensure the boat is protected with good drains before the trip is started.

Hatches & Portlights

Deck hatches can be another source of leaks, but fortunately these can usually be tackled successfully without major surgery or expense. Hatches usually leak due to their seals being perished or damaged. Check they have a reliable method of being locked from below. In bad cases where leaking is difficult to stop have a canvas cover made for heavy weather. I carry a roll of heavy duty brown parcel tape: which can work wonders if applied around the hatch edges, but of course the hatch has to dry, otherwise the adhesive will not stick effectively.

Opening portlights in the coachroof are another source of annoying leaks. Ensure the fit is good and if necessary re-bed with a flexible compound or fit new seals. If your boat is fitted with opening portlights in the hull, lock them up solid before the cruise starts and leave them closed until it's completed. In my view no offshore yacht should have opening ports in the hull sides, for it's almost a certainty that someone will leave one open sometime with potentially disastrous results.

If your boat has large windows say over two feet square then it's a sensible plan to have a set of stormboards made up of three quarter inch ply for use in heavy weather. Ensure the fixings are strong and can be easily done up, otherwise, human nature being what it is, they will never be fitted. Dorade vents can, and do leak at times, so make sure they can be closed off completely from the inside when conditions get rough.

Other areas of potential leakage are around the mast, through deck fittings such as genoa tracks, cleats and electrical plugs. These are likely to be irritating rather than dangerous, but are certainly well worthwhile checking over, before departure.

The other main hole in the hull, is the engine drive unit. In the case of prop shafts the stuffing box will leak if it is not correctly maintained.

The stern gland needs periodic adjustment to keep it watertight. If the stern tube is found to be leaking, tighten the greaser (if fitted). If this does not cure the flow then the internal packing will need compressing. This is done by hardening up the flange a little. Usually the stuffing box has two bolts on the inboard end, these need to be tightened up – about a quarter of a turn at a time until the water stops. Be careful not to over-tighten the gland because it can lead to the shaft overheating. If no greaser is fitted, the valve gland should then be dismantled when the boat is laid up out of the water, and repacked with grease.

If there is no remote greaser fitted it is sensible plan to put one in. They are relatively straightforward to install and the actual greaser unit can be remotely attached to any convenient part within the engine compartment. If your boat has a P-bracket supporting the shaft outside the hull, inspect the cutless bearing. This bearing is water lubricated and if there is any appreciable play within the P-bracket housing when moving the shaft up and down, the bearing is due for replacement. In recent years there have been a number of internal shaft seals marketed. These replace the stuffing box completely and are totally maintenance free. Their reputation is impressive and we have had one for some years now. A point to note nevertheless is that unlike the stuffing box they do not support the shaft in any way, so if the shaft is a long one, an extra shaft support bearing may have to be introduced, to avoid 'shaft whip'. The manufacturers will give advice on this.

With an engine sail-drive unit things are less complicated. The main concern is the large hole in the hull from the actual engine installation. The water seal around the unit, is made of tough rubber, but it does deteriorate over a period of years, and begin to perish. The concept of these drive units is still relatively new, so the life of the membrane seal cannot be predicted with certainty. However some units seven years old are still showing no signs of leakage so it's a reasonable assumption that a good five to six years life expectancy can be relied on as a minimum before replacement is likely. The first signs of the membrane starting to perish will be water seepage around the engine bed, so check this when the boat is afloat. It is unlikely the membrane will suddenly collapse; it should just start to weep, this is the sign it's deteriorating. In any case just to be sure, I would adopt a five year routine for replacement.

Bilge Pumps

There is an old saying that the best bilge pump is 'a frightened man with a bucket'. This may be true but it goes without saying, that every seaworthy yacht must have at least two reliable bilge pumps, fitted in positions where they can be operated efficiently. The suction ends should be fitted with strum-boxes to filter out any debris which may be in the bilges. This will ensure the pumps do not get blocked. When siting the pumps, give thought to the operator, who may have to be at pumping stations for some time. It is worthwhile considering extending the standard levers (which always seem to be short) to give better leverage so that the pumps can be used to their maximum potential. There are many excellent pumps on the market, some single, and some double stroke models. The best ones in my experience are the diaphragm types such as Henderson or Whale which can be quickly dismantled if they become blocked.

Siting the pump inlets is also very important. In keel yachts with a deep bilge there is no problem, just get the suction as low as possible. With shallow bodied yachts with little or no bilge or pump well, even small amounts of water can be a real nuisance. The answer lies in not siting the inlet on the centre line only, but to both sides of the bilge via a simple change over valve. In this way the yacht can be pumped out when upright and when sailing heeled on either tack.

On *Mishka* we have an added problem. Although she has two bilges, they are shallow and separated fore and aft by the built in water tanks that run athwartships across the main cabin. This effectively cuts off the bilges from each other, so in order to keep her dry, we have two separate systems plumbed in. This is a nuisance that should have been solved at the design stage, by incorporating a wide mouth limber pipe through or around the tanks leading to the main bilge aft.

Electrical pumps now, are well worth installing, as they can move considerable quantities of water fast. In my estimation, they should be considered, but with the back up of a manual system, just in case all power is lost in an emergency. The fully submersible types are the best and quite inexpensive. On some boats they are fitted with automatic floats to activate the pumps when the bilge water reaches a pre-set level. This can be a very useful safety aid, but should not be relied on totally as with everything mechanical they can go wrong.

The main engine can also be used as a powerful, tireless pump in an emergency. Disconnect the water inlet hose from its seacock (Shutting the seacock first!) and place it in the bilge, but make sure it is fitted with an efficient strainer, for if any debris is sucked in, it will quickly damage the water pump rendering it and the engine inoperative.

The head can also be pressed into service as a bilge pump. There is a simple device on the market which plugs into the bowl outlet, then the attached hose can be led to the bilges for removal of water, utilizing the powerful head's pump. I have not tried it myself but it looks practical and maybe worth trying. The main thing is to have a sufficient number of reliable pumps available to move quantities of water quickly. Finding a leak when the boat is

rolling and pitching is never easy, and if the cabin is flooded as well, it makes things even more difficult and nerve racking.

Working the Deck and Cockpit.

Deck work is potentially the most dangerous aspect of offshore sailing and as such should be taken seriously by anyone contemplating a blue-water cruise.

The main priorities are, firstly to make the decks safe and convenient to work on, then to ensure all the deck gear is strong, correctly positioned, and in first class working order.

With many modern boats the sail handling gear is led aft where it can be operated within the comparative safety of the cockpit. This is a great advance and should be applauded. There are however very many older yachts that don't have this convenience, which means the trip to the foredeck has to be made frequently and sometimes in conditions that are unpleasant and dangerous.

The Deck

Secure footing is a vital component for safe working of the ship. It is an unfortunate fact that many moulded non-slip patterns on GRP cruisers are ineffective, in particular, the ones found on coachroof tops seems to be the worst. A good hard look should be taken at the decks and coachroof. If at all doubtful about the non-slip qualities on your particular boat, consider it a priority to do something about it. The most vulnerable areas are immediately adjacent to the coamings where the crew step from the cockpit. The cockpit seats and sole should be beyond reproach as should the deck or coachroof areas around the mast.

Plastic deck hatches are notorious skid pans when wet and they should receive special attention. Always remember that a boat is never still and can have a heel of up to 30°, so when visualizing areas of support for going forward, the coachroof side can become a useful level foothold!

A balance must be struck when contemplating improving on non-slip surfaces, between secure footing and a harsh surface that will damage clothes and abrade skin. If the deck is hopelessly smooth, just simply paint it with non-slip deck paint, this can be very effective and lasts surprisingly well. Individual small areas such as the ones mentioned above can be made secure by applying Trackmark™ or strips of self-adhesive 3M™ non-slip tape. The latter is particularly effective on hatches as it does not obscure too much light if the strips are narrow.

Stanchions & Guardwires

Guardwires should be double, and run along the whole length of the boat. Where they attach to the pulpit and pushpit the connections should ideally be via lanyards. If bottle screws are used, ensure that there are no metal to metal connections. If this cannot be avoided use an insulator at the point of attachment to break the electric loop that interferes with radio signals.

A height of twenty eight inches is the ideal for stanchions, but on small boats this may be a little high and look out of proportion. In this case use stanchions of twenty four inches, which is a good compromise and is the standard most modern boat builders apply.

It is most important that the method of fixing the stanchions to the deck is really strong. Through bolts to the underdeck are vital, and these should be backed up with half inch plywood pads to spread the load. It is also worth considering mounting the stanchion feet onto wooden pads at deck level, to minimise shock loads on to the deck surface should a heavy crew member fall into the guardwires or stanchions.

Contrary to some people's views who consider stanchions and guardwires are mainly of psychological benefit, I believe they have a serious role to play and provide a real barrier to falling or slipping over board. Provided of course they are securely attached to the hull.

Jackstays and Harnesses

Deck jackstays must be considered essential equipment for all offshore yachts no matter the size or type of boat. The lines should run fore and aft both sides of the deck or coachroof and be long enough to enable a crew member to reach the boat's extremities whilst clipped on by his harness. It is important not to stretch them too tight, but to fit them with approximately six inches of play when taken from the centre point. Make up the stays from stainless steel wire and have the ends finished with a talurit terminal or swaged fitting.

Ensure the U bolt fixing is through the deck using quarter inch stainless steel bolts then backed up with half inch plywood or quarter inch alloy sheet pads. Don't skimp on this as it could be the very fitting that may save a life. The shock loads of a heavy man falling overboard whilst attached to the jackstay are considerable, I have seen a poorly attached fitting just pull through the deck like a rotten tooth! I believe jackstays should be stainless steel wire even though the wire can roll underfoot at times; if the crew are aware they are there they should not cause much of a problem. To improve on wire, some boats of late have started to fit jackstays made from one inch flat webbing. I am not very keen on this as UV degeneration is accelerated in areas where the sun shines for long periods. If the decision to use webbing rather then stainless steel wire for jackstays has been made, ensure it is replaced regularly at no more than eighteen month to two year intervals.

Harnesses

A full harness should be carried for every crew member on board, and a harness routine adopted that is religiously carried out by all members. It can be a bore to put on harnesses when the weather is mild, with light winds and calm seas but this can be just the time when crew members are off guard and at their most vulnerable. The strict harness discipline we have adopted on our boats when offshore, is simple and reasonable, so it has a far greater chance of being carried out. Basically, harnesses are always worn and *clipped* on when entering the cockpit at night. They are used in conjunction with the jackstays by everyone who goes on deck, day or night. In daylight hours they are not used if the crew stays in the cockpit. If the weather is above Force 5 and there is a chance that the sails might need reefing, harnesses are normally worn. We have a harness shelf by the companion-way just to one side of the main steps, so that they are always to hand. The cockpit has strong U bolts for harnesses sited within easy reach of the companion-way steps, so it's a simple matter to clip on before stepping into the cockpit. The helm also has extra harness points in order that he/she can be securely clipped on whilst steering the boat.

I prefer my harnesses to have quick release hooks at both ends, plus an extra hook about a third way along. This enables the harness to be used on a long scope or short one and gives the wearer greater flexibility when working around the mast or on the foredeck. The new Gibb type hooks that cannot be accidently tripped are much to be preferred to the old type of carbine hook.

Some foul weather jackets have very good harnesses built in, these have much to recommend them but of course harnesses are not always worn with oilskins, particularly in the warm Mediterranean or tropics, so individual harnesses should always be available on board for the crew, whatever the prevailing weather conditions.

Caring For The Sails

Sails are the primary moving force in ocean cruising, so it makes sense to treat them with the care their importance warrants. Tropical and Mediterranean conditions that give long hours of sunshine may be delightful to us, but it's murder for our sails. Strong ultraviolet rays are surprisingly destructive to modern sailcloth and it pays handsomely to cover a yachts sails at every opportunity, when they are not being used.

The most used and abused sail on any yacht is its mainsail, for it's up in all conditions, reefed or unreefed. It is vital therefore that the condition of this sail is beyond reproach. I would recommend that before anyone embarks on a long distance cruise the mainsail is appraised critically. If the existing sail is over seven years old, replace it for a new one. If the age and general condition is good, then send it to a sailmaker to have it carefully checked over. Get them to pay particular attention to the high stress and wear points, clew, tack, headboard, slide fastenings, batten pockets and areas that bear on crosstrees and pulpits etc. Then have it triple stitched along all the seams. Check and strengthen if wear is apparent around batten pockets and reefing cringles. Ship a couple of long battens as spares (these can be cut down to fit the smaller pockets if required) plus a supply of sailcloth for emergency repairs. I have found the sail material used for sail numbers is ideal, it is self adhesive with easy self peeling backing. It comes in white as well as a number of other colours and costs very little. A small roll in the bosun's locker is well worth its keep.

Contrary to some people's belief it's not always blowing a gale in the ocean. Quite the contrary in fact. With blue-water sailing, gales will have to be met and managed of course, but provided routes are selected at the correct times of the year modest wind speeds should be the norm.

Light or moderate winds will predominate on a trip to the Mediterranean from late May to early September so it is most important that light weather sails are shipped aboard.

There is nothing to beat a large ballooner, built of light material to get a boat moving. It's best tacked down to the stem head and set flying. It will pull like a horse from about 30° off the wind. We have a flat cut cruising chute of large area for this purpose, and when combined with the mainsail and mizzen we can sail close to True Wind speed in light airs.

Spinnakers have their place in long distance cruising, with large crews but particular attention must be paid to their gear. By their very nature spinnakers are restless sails that cause high loadings on the rig. Chafe is the biggest enemy, so before commencing a long cruise ensure all the leads are fair and high chafe points well padded. Replace any suspect running rigging, sheets or guys that show wear. Strips of light towelling bound and taped round pulpits, guard-rails and terminals are extremely effective. They may not look particularly attractive but can be quickly removed once the main cruise is over.

Other sails such as genoas, working jibs, mizzens etc. should also be carefully checked over, washed and repaired as necessary. Remember, it's much easier to get these tasks done before you leave, rather than in the West Indies or the Mediterranean where facilities are limited and expensive.

Keeping Sails in Good Order

No boat should venture far offshore without a full sail repair kit or sewing machine. You should also have an adequate supply of spare sailcloth in various weights that correspond to the sails on board. Simple tears and repair patching can be carried out by the most inept of us males, but larger repairs should be left to professional sailmakers to tackle. It is worthwhile however to do a little practice with the needle before you leave, so at least the repair can be started with a modicum of confidence.

I also carry a roll of spinnaker repair tape for small tears.

Sail repair kits can be purchased from most good chandlers but I suggest the following is a sensible minimum:

Various thicknesses of twine and thread for stitching.

A good quality sewing palm.

Beeswax for lubricating thread.

Knife.

Assorted sail needles in waterproof wallet or container (ensure they are packed in VPI paper to inhibit corrosion).

Small quantity of webbing for repair of slide attachments.

Quantity of spare mainsail slides.

Hanks for foresails, if roller furling is not fitted.

Chafe is the most common damage sails receive at sea so it is sensible to appraise the rig critically to reduce this danger to a minimum. It is far preferable to prevent rather than cure possible damage.

Mast Ladders, etc.

Fixed mast ladders, although valuable for checking the rig aloft are a frequent cause of chafe. Examine them carefully and remove any sharp edges you find. Bottle screws, and lifeline attachments may have split pins, so make sure they are well taped. Carefully examine the spreader ends, replace any damaged 'boots' and apply plenty of extra tape so that they are really smooth and will not damage the genoa. It is well worthwhile having chafe patches put on by the sailmaker before you leave, on areas where the sail rubs across the spreader ends, pulpit etc.

Maintenance

Look over your sails very critically before you leave. If the nylon mast slides show excessive wear, replace them with new, they are very inexpensive. I much prefer fixing sail slides with webbing tape but probably this is best left for the sailmaker to carry out.

There is no doubt that the ability to service one's own mast is most convenient, therefore some form of mast ladder is worth having. That illustrated is not permanent, which although extravagant on storage space is very safe in use.

Early mild weather in the Med generally means that fitting out can begin in March.

Chapter 8
Engines and Associated Worries

Engines, Maintenance, Oil, Filters & Additives, Engine Spares.

The reliability of the modern diesel engine is now an accepted fact of boating life. Yet it's not that long ago when yachtsmen had to struggle with temperamental petrol models, that were covered in rust and lurking in a damp dingy space in the bilges. The breakthrough came when it was found that diesels could be manufactured of lighter materials. Modern technology was then used to give us higher power to weight ratios. This means that even very small yachts can now have a small powerful unit installed without upsetting the boats trim.

For Mediterranean sailing I believe an engine is a virtuall necessity. There can be long periods of calm weather to cope with and frequently strong winds and high seas also, so some form of power is essential if even modest schedules are to be met. A powerful engine makes sense too, for the seas in most parts of the Mediterranean are choppy. The local winds can blow strongly at times and it is very useful to have the extra power to push against the elements to make harbour at the end of the day. On passage, a powerful engine will also pay its way. Under summer conditions it is not unusual to have calm nights with little or no breeze for sailing. To make reasonable progress on passage therefore, motoring will be necessary. If the boat can motor close to hull speed, then, it will enable the yacht to keep up good averages.

I took my own advice when re-engineering our boat and gave her an 80hp unit. This is a little overkill perhaps, but it does enable us to cruise at seven and a half to eight knots under most conditions. An added benefit is that with all that power available she manoeuvres like a tugboat.

The trend over the past few years is for manufacturers to offer uprated engines by turbo-charging their normally aspirated models. This substantial increase in power is happily not at the expense of weight, so from the yachtsman's point of view it is a bonus. Nothing is for nothing however, and there is a downside. Firstly, to obtain the extra power, turbo charged engines have to be run at high revolutions to bring the turbo charging into effect. This increases noise, wear and fuel consumption. It is not uncommon for these units to peak at 3,500 and even 4,000 RPM, so there is a lot of thrashing going on. Secondly, turbo engines should not be run at idle or on light loads for long periods, as the turbo units tend to carbon- up and glaze the engine bores. This will ultimately harm the engine and result in major repairs. Along with many other yachtsmen I occasionally enjoy just pottering along with the engine ticking over slowly. This will not harm the engine for short periods, but to keep the engine happy, I then run it hard for ten minutes or so to get it really hot and blow out all the deposits. Diesels thrive on hard work and it's much better for them to attain their working temperature quickly. For battery charging in port, try not to leave the engine idling in neutral for long periods. Tie up securely to a pontoon or wall, put the engine in gear and run it gently under load. It will use little extra in the way of fuel but it will be much better for the engine.

Maintenance

If the engine in your boat is an old one, give it a birthday before you leave and have it rebuilt. Although this will not be cheap, the first cost will be the smallest, for to get engines repaired in the tropics or

Mediterranean is sometimes difficult, due to the lack of spares and skilled labour.

It is worth putting the owner's manual on board for your particular engine. Don't confuse this with the owner's handbook. The manual sets out all the engines parts and gives each one a number. It will also be available for a mechanic to enable him to carry out repairs if the engine breaks down in some out of the way place, where only local and relatively unskilled labour is available.

Some manufacturers run training sessions for owners. This really does make sense for it's a comfort to understand the workings of ones own engine and be able to carry out the necessary maintenance oneself. Not only will this ensure the work is done on time, but also that it's carried out properly. At the very least the engine should be completely serviced before the trip is started. Routine maintenance is not difficult. It basically calls for changing the oil and filters and bleeding the fuel system, tensioning the fan belt etc. All these tasks should be well within the capabilities of the average owner. If an engine starts out in good condition and is maintained regularly, it should give very little trouble and be completely reliable.

Oil

The engines life blood is its lubricating oil. A major part of the wear in a diesel engine occurs on starting and when warming up. In diesel fuels, there are minute traces of sulphur, and when this is combined with water (also found in fuel) it forms a solution of sulphic acid, which quite obviously is harmful to bearings and engine surfaces. If the engine is run at low speeds or under light loads, it will not reach its operating temperature and in consequence, piston rings and cylinder wall wear will be accelerated. Carbon is also washed down the cylinder walls to pollute the oil further and if the old oil is left for long periods

without being changed it will form a sludge that may block small oil passages, with disastrous results for the engine.

One of the problems associated with small high revving diesels is that their oil sumps tend to be on the small side, therefore less oil is being asked to do more work! It is therefore vital for the long term health of an engine to change its lubricating oil and filters frequently. This should be done at least to the manufacturer's recommendations, and if the engine is used only infrequently under light loadings or operated in areas where the fuel is dirty, an extra change each year will not come amiss. Always change the oil and filter at the end of the season in any event.

Changing the oil can be a messy business unless one is well organised. To start with, run the engine until the oil is hot, this thins it down and ensure the majority can be removed. Obtain a special sump oil pump. This is a simple unit comprising a plunger pump with an outlet hose and thin metal tube for inserting into the engines dip stick hole. The hot oil is then simply pumped out into a convenient receptacle. Make sure this can or bottle has a larger capacity that the total of the oil being removed from the engine, also use a container that has a screw top or lid that can be firmly closed over the old oil to avoid spillage. If you find the pump tube is too short to reach the bottom of the engines sump, replace it with a short length of plastic hose and clamp this securely over the engines dip stick hole with a jubilee clip.

Once all the oil is removed replace the engine oil filter. These are usually unscrewed in an anti-clockwise direction. If it's difficult to remove, drive a screw driver right through the filter at the outboard end, and use it as a lever. Place a plastic bag over the complete filter and remove, being careful not to allow any oil that seeps from the engine to drip into the bilges. The bag can then be sealed and disposed of in the usual

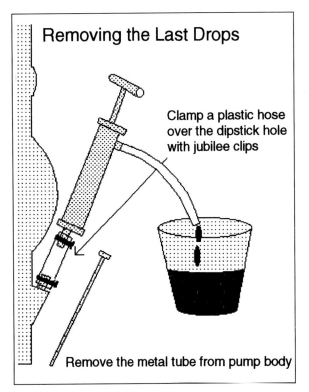

Removing the Last Drops

Clamp a plastic hose over the dipstick hole with jubilee clips

Remove the metal tube from pump body

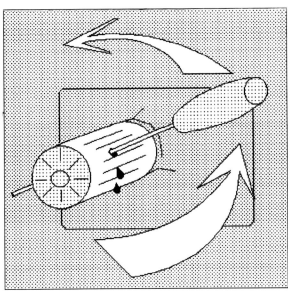

Sometimes it is difficult to remove the oil filter. This is best done by carefully driving a large screwdriver through both sides of the filter and then using the increased leverage to unscrew it in an anti-clockwise direction.

Getting all the old engine oil from the sump is not always possible if the pump tube is too short to reach the bottom or too thick to fit in the dip-stick hole.

A simple way round this is to remove the tube and clamp in its place a short length of plastic hose.

way. If the operation is carried out with care, there is no reason why it should be messy. If it's not, the mess ensued by quantities of black glutinous oil is indescribable. I have a plastic bag nearby for dropping the oil pump into immediately it has been used, so the possibility of drips from this finding its way into the accommodation is also eliminated.

Fuel and Filters

The fuel filter system used on many yacht auxiliaries is based on a primary filter close to the fuel tank plus one other on the engine itself. Usually the primary filter is of the water separation type, in that it has a glass bowl under the main fuel filter. It is there to collect any water that is in the fuel. It can be drained off by turning a small tap in its base, and as it does not interfere with the main fuel supply to the engine, there is no need to bleed the system afterwards. The fuel filter element above the bowl collects any sludge or small debris in the fuel, so the element will need replacing once it becomes clogged, and before it starts to interfere with the fuel flow to the engine. The secondary filter fitted to the engine is usually of a finer type and is the back up to the primary filter. It is necessary to change these filters regularly, and at least once a

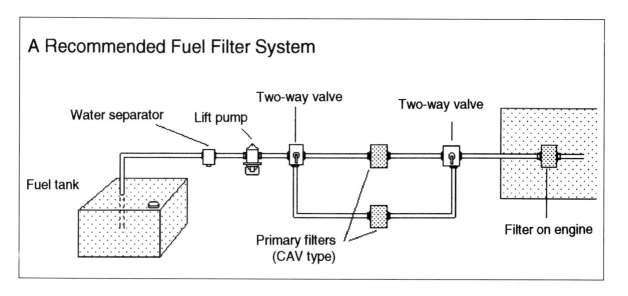

A Recommended Fuel Filter System

Water separator
Lift pump
Two-way valve
Two-way valve
Fuel tank
Primary filters
(CAV type)
Filter on engine

By using two separate fuel lines, each with its own primary filter and connected by two-way valves, the fuel can be switched to a clear line if a blockage occurs without stopping the engine and bleeding the system.

year, in order that the fuel is kept as clean as possible.

The injectors spray a fine mist of fuel into the cylinder in a pre-determined pattern and as such are precise instruments that need servicing every thousand hours or so. If they do become blocked or damaged in any way the engine will not function, so clean fuel is the number one priority for smooth trouble-free engine running. Unfortunately it is very difficult to obtain diesel fuel that is completely free from all water contamination, especially in the out of the way places in the Mediterranean or Caribbean. Another source of water in the fuel is from leaking deck filler caps or from condensation in half empty fuel tanks left over the winter period. On laying up, or if the boat is left unused for some months, top up the tank before leaving.

Fuel Additives

For sometime now, we have used fuel additives in our diesel to combat the water pollution that invariably seems to be present. Research has been carried out at the Thermo-Fluids Engineering Research Centre, City University, London, into this subject. Products have now been evolved that blend the water and fuel into a solution that will burn cleanly within the engine, without any adverse effects. These products are economical to buy and also seem to make the engine run more efficiently.

Spares to Take

Availability of engine parts is a problem when cruising away from home waters. It's important to be as self-sufficient as possible. If the cruising schedule enables regular visits to home ground then the problem is less acute, but if you envisage being out of touch with agents for the particular engine fitted, then the spares list should be comprehensive.

As a guide the following is the minimum engine spare list that should be carried:
    1 fan belt.
    1 water pump and impeller.

4 of each of the fuel filters fitted.

4 oil filters.

2 cans of engine oil for two complete engine changes.

1 injector.

1 can gearbox oil.

1 thermo-stat.

1 engine manual.

Spare engine cooling hoses and jubilee clips to fit.

Oil

Only use the best quality engine oil from recognised manufacturers that has been specially blended for diesel engines. Choose a viscosity rating suitable for your cruising area. There are many multigrade oils on the market that are suitable, but as a guide SAE 10 is for cold weather, SAE 40 is for hot climates. The American Petroleum Institute (API) has developed a classification that is used on oil sold in many countries including the Caribbean. It is rather complex, but when buying oil look out for 'C' which means it is manufactured for use in diesel engines. This is followed by another letter which basically indicates quality i.e. C, D or E. Therefore CE is the best available at this moment. Further classifications will be added in the future. Watch out for CX planned for introduction in 1994.

Chapter 9

The Power Game

Electricity, Batteries, Charging, Wind & Portable Generators, Solar Panels.

It is quite staggering to add up what electricity the average modern cruiser uses in normal day to day living. With ever increasing demands on the energy supplies brought about by sophisticated electronic wizardry being bolted onto the yachts electrical system, it's no wonder that many yachts find their standard batteries just cannot cope.

Firstly, let us examine batteries themselves and try to calculate the maximum energy supplies a yacht will need for cruising in the sun.

Battery Types

There are three main types of lead acid batteries available:

1. *Automotive.* These are the familiar batteries that are fitted to our cars. They provide a high starting current and will accept recharge quickly.

2. *Storage.* The advantage with this type is that they can be heavily discharged slowly, with little damage to their life. They are not suitable for engine starting.

3. *Deep Discharge.* A heavy duty battery that combines the advantage of both the above. Because of their heavier build standards, they are the most expensive of the three.

In addition to these basic types, there are other choices:

1. *Normal Vented types.* These have removable tops to enable electrolyte to be added when required.

2. *Gel batteries.* These are batteries that have a chemical added to the electrolyte to

make it into a gel. This allows the battery to continue to operate even if it's inverted for a short period.

3. *Sealed types.* These are sealed units, water or gel, and are sealed against spillage. They are of the automotive type and are not available at the present time as deep discharge batteries.

Every yacht should have two separate power banks. One dedicated to engine starting, which can be of the automobile type and the second bank for domestic use; the deep cycle type should be used here, as the discharge rate is likely to be slower and with lighter amperage drawn at any one time.

It is important not to mix the types or classes together. If the choice is to use vented batteries they should all be the same. Do not connect gel and vented types in parallel or in series.

How much power?

There is some confusion as to how much storage is really needed. If one looks at the standard equipment used on many boats nowadays the answer must be considerable. I am often asked why is it that boat batteries don't seem to last as long as the ones fitted to our cars. The answer lies in the use these batteries are put to. The standard auto battery is just used to start up the car in the mornings, when it gets a hefty discharge for a second or two, the power is then replaced by the engines alternator as the car is driven. Under this type of treatment an auto type battery will stand many thousands of charge/discharge cycles and still keep itself in good shape. With a boat's battery, the treatment it is subjected to is very different. The discharge is gradual, and as it comes from a number of sources it can go unnoticed until the discharge level falls below fifty percent. This can happen on a night passage for example – or if the autopilot and refrigerator are used non-stop over a longish period without

running the engine. This complete discharge can do a great deal of harm to a battery, and if repeated frequently it will ruin the battery completely within a comparatively few charge/discharge cycles, perhaps even as few as fifty. It is therefore critical for the long term health of a battery to examine the way it's charged and discharged, and never let if fall below fifty percent of its total capacity.

Whilst on the subject it may be worth examining just what power is available from an individual battery.

A standard battery may be rated at say one hundred and twenty Amp Hours (AH) when it is assumed it will be discharged over twenty hours. If current drawn is over a shorter period, say ten hours or less, the available power is reduced by up to twenty percent, so, the higher the rate of discharge the less efficient the battery is likely to be and the lower AH rating is available for consumption. All this assumes that the battery is capable of holding a hundred percent state of charge(SOC) in the first place, which is not always the case, especially as the batteries age. Their holding capacity will be reduced if they have been neglected or abused, so in practise the total charge may well be eighty percent SOC or less and not the hundred percent assumed. Another important point when considering charging, is that a battery that has been discharged slowly will resist being recharged quickly.

The engine alternator is an efficient method of replacing energy that has been taken out of the battery quickly but inefficient at replacing energy that has been discharged slowly. Under these circumstances, only a shore based trickle type charger can restore a battery to its full hundred percent SOC state. The reason for this is that a battery quickly builds up resistance to being recharged and the alternator cannot uprate itself to cope with this added resistance; unless modified by the addition of an electronic regulator such as the TWC or similar.

Standing batteries will also lose power gradually through natural drainage and through low loads from clocks and leakages in the circuit wiring etc. Always try to replace lost voltage gradually and try to leave them fully charged when the boat is not used for a period of weeks.

It is a good thing to remember that if a battery is discharged more than fifty percent of its state of charge, it will be damaged. Close monitoring of the battery system is essential and the purchase of an electronic digital battery monitor, such as a Smart Power Monitor or similar, is a useful aid. These will monitor and remember the current that passes back and forth and in addition will keep a cumulative record. They will show on a display present consumption, alternating with cumulative AMP/HRS put in or removed from the battery.

Putting the power back in

The simplest and best system of replacing power is to hook up the batteries to shore power via a 220 volt charger. There are many models on the market varying in output from around 10 amps to 40 amps or more. The more sophisticated units will have built in systems to 'float', so that the current will be switched off automatically when the batteries are fully charged. Most marinas in the Mediterranean have this power facility available for each berth and the running costs are usually included in the overall berthing charge.

Engine Charging

By using a 'bolt on' electronic charging regulator that uses micro chip technology most batteries can be recharged to a hundred percent of available SOC. These clever little black boxes achieve this by side stepping the counter-voltage resistance in the batteries. They set up high voltage

bursts and intersperse these with lower rates of charge, whilst the 'microchips' measure the voltage in the battery. All this actually goes on whilst the engine is running in the normal way, and it is claimed that these units will extend a batteries life by up to two or three times.

When electricity is required to power such things as freezers, air-conditioning units, microwave ovens, water making plants and the like, power generation enters another league. It will then be necessary to look elsewhere than the engine to provide enough power to keep all these appliances going when offshore. The sensible answer is the built-in generator, which can be used on 'automatic' to cut in when the batteries fall below a pre-determined level.

On a less demanding plane, and the majority of boats really fall into this category, power output away from the shore-based unit can be maintained in a number of ways.

Solar Panels

For Mediterranean and tropical sailing where the sun shines for up to sixteen hours a day in summer, the solar panel can be a most useful way to supplement the engine. These solar panels are superb for silent trouble-free power supply, but their individual output is low, so must be considered as the 'offshore trickle charger' unless a bank of three or more are installed. They do take up rather a lot of room but are well worth this inconvenience, especially if the boat is left for long periods away from shore charging facilities. The panels can be wired up to a regulator that cuts off the power when the batteries are fully charged. This enables the panels to be 'left on' permanently when the boat is unattended for lengthy periods, without the risk of overcharging taking place. In my view they are ideal for blue-water cruising.

This Aquair model from Lumic Ltd, is extremely versatile in that the generator from the wind unit can, with the addition of a towed impeller, be turned into an efficient water turbine. This is ideal for long distance sailing under down wind conditions, where the wind units tend to be least efficient.

Wind & Water Generators

Another reliable method of producing power is by wind generation. There are a number of excellent wind generation units on the market that have proved to be very reliable. Their output is geared to actual windspeed and the best ones seem to charge at around four amps in eighteen knots of wind. One disadvantage with these windmills is that when the yacht is running, Apparent Wind is considerably reduced, so the charging rate drops off dramatically. This is a particular problem with long distance cruising, say from northern Europe to the Caribbean, when one can reasonably expect the majority of winds to blow from astern. Once the yacht has arrived however, a good unit should be able to keep the batteries topped up without too much trouble. We have given this matter some though and for *Mishka's* crossing will use a combination of a towed turbine when at sea and a wind generator when at anchor. The towed turbine is claimed to have a rather good performance, being quoted as five amps at six knots through the water. The advantage with this particular system, the Aquair 100™, is that when at anchor the water turbine is unhooked and the generator simply hoisted in the rigging with a six blade wind turbine attached, which turns it into an efficient wind generator.

The towed water turbine comes in two weights and sizes, the choice of which depends on the speed of the yacht. The standard unit at twenty pounds is used up to seven knots and the larger, course pitch model is used for sustained speeds of seven knots and more. Drag is always a concern for sailing yachts, but the quoted figure of approximately half a knot is a price I feel worth paying in return for fully charged batteries. The downside of this system must be that the air unit is not fixed to the yacht, so there will always be some setting up time whenever it is used.

There are several other excellent permanently fixed air units available which provide good performance. These can be mounted anywhere on the yacht, but of course do not have the advantage of the towed power plant at sea. The wind generator is probably not as ideal for the Mediterranean as it is for the Caribbean, since the winds do not blow so reliably.

Portable Generators

The small petrol driven generators also have their place. We have one, and on several occasions it has been a godsend. It has enabled us to get out of trouble when the batteries have run flat. They will in fact put charge into a completely flat battery which is not always the case with other charging systems. Due to their high fuel consumption however, I put them into the emergency category rather than utilizing them on a regular basis. They will of course power other things such as fan heaters, power drills, electric lights and for the girls, most importantly, hair dryers! We have 'plumbed' our model in, so that it is in fact immobile with the exhaust led through the transom. This seems to work quite well but we do take great care not to spill any petrol when refuelling, and we pay particular attention to ventilating the locker when it is running. They are really designed to be used in the open air but can be fixed 'in situ' if the locker where they are mounted is sealed off completely from the accommodation, with the exhaust led over the side. A housing box mounted on deck would be ideal.

Chapter 10
Is She Ready for Cruising to the Sun?

Keeping Cool Below Decks, Ventilation, Cockpit Tables & Shading, Awnings, Refrigeration.

Keeping the Crew Cool

It may seem strange to concern ourselves with keeping cool when the purpose of this book is to encourage yachtsmen to cruise to warm climates. The fact remains however that without the ability to remain cool by choice, cruising in the sun would quickly become a misery.

A constant movement of fresh air throughout the boat is essential for comfortable life aboard. The boat quickly heats up with the arrival of the sun each morning and without adequate ventilation temperatures can soon become unbearable. One of the disadvantages with some modern yacht designs is that they have large windows to make their interiors light and airy. This is fine for cool climates but, impossible for hot ones. Adequate curtaining must be provided in order that the sun is unable to penetrate the interior. It is no coincidence that houses in countries that enjoy hot weather over long periods, are designed with small windows which almost

A well designed windscoop will catch even the slightest breeze and channel a welcome draught of air below to cool the boat.

Windscoops

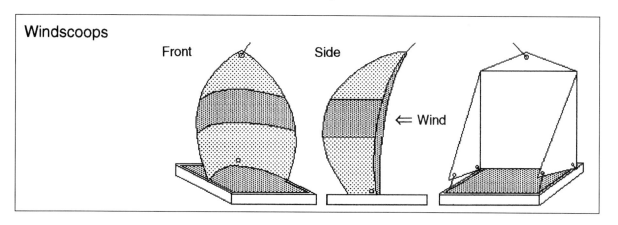

There are many windscoop designs available, but a simple one can be made up using soft light fabric to fit individual hatches. It is important that sufficient eyelets are put in, so that they can be tied down securely to avoid them being blown away or flapping too much.

invariably have shutters that can be closed during the heat of the day.

Hatches should be large, and where possible provided with windscoops. Sleeping at night is sometimes difficult in high summer; then the only option is to use the cockpit, when some form of overhead shelter must be rigged to protect the sleepers from the heavy dews that are a feature of Mediterranean weather.

The Caribbean is rather better from the point of view of keeping cool. The trade winds blow almost without pause for up to ten months of the year, so it is a relatively simple task to arrange for cooling breezes to blow through the yacht by fixing up wind scoops of one sort or another. As the breezes can be very strong at times, a half gale can blow down below as well, so some form of 'reefing system' for the scoop is advisable. We also use electrical fans in the aft cabins which sometimes get very stuffy. They stir up the air, but can only be used for relatively short periods for they tend to be rather noisy and heavy on electricity.

The Cockpit

The ideal cockpit should have side benches at least six and a half feet long to accommodate a full length sleeper. I love to sleep on deck in this way, and watch the dawn break, with all the noises of the wildlife awakening to a new day, it adds another dimension to cruising enjoyment for me.

In a warm climate, the cockpit is the area most used by the crew and as such should be made as comfortable as possible. We have most of our meals here, even dinner, and when it's dark we rig a small lantern from the rigging to provide sufficient light. Have some tailored cockpit cushions made up to fit all the cockpit surfaces, but keep the thickness down to two inches, which is perfectly adequate for most posteriors and takes up less room than the thicker ones. It's also sensible to take two sets of fitted towelling covers for all the cushions, made up in light colours to reflect the sun. These overcovers are very necessary as the original cushions will quickly become stained and sticky with suntan oil and the like. Towelling is very easy to wash and it can be purchased in long lengths from any major departmental store.

Cockpit Tables

As most of the meals will be eaten in the cockpit a table is necessary for comfortable eating. Some production yachts now have one fitted as standard which is very civilised,

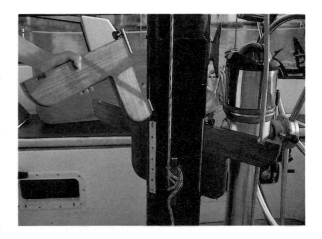

Cockpit tables are a wonderful convenience that is essential for warm weather cruising.
On Mishka we have used the mizzen mast as the table support with removable slide-in supports(Top). Once erected the table takes little room but has ample space for four people to sit around.

but all boats can be rigged with a demountable table if a little though is given to its design. If wheel steered, use the binnacle as the centre support. On *Mishka* we have this and a mizzen mast, so we have designed a table that is very solid yet quick to rig. It is stowed away on a specially made shelf in the lazarette and held against the bulkhead with shock cord. In a yacht that is tiller steered, a free standing table is the best option. On a previous boat we owned, we designed a simple cockpit table with removable legs. These were purchased at a DIY store and screwed into base plates that were attached to the underside of the table. It worked very well for a number of years, the only problem was that it was rather bulky to stow below.

Shading the Cockpit

Some sort of removable awning for the cockpit is a necessity if long periods are to be spent on board. Ideally it should be of the Bimini type and simple to rig. The advantage of a Bimini over an awning spread over the boom is that it can stay up whilst sailing, which doubles its usefulness. If some sort of shade is not available, then the crew will gradually fry in the sun. This is not an over statement, for it is just as important to be able to keep out of the sun, as is the sunshine itself.

The dangers of over exposure to ultraviolet rays is now becoming understood and accepted as one of the prime causes of skin cancer. Sunshine is a wonderful tonic and uplifter of the spirits but it does have its dangers. It is prudent if sailing in the tropics or the Mediterranean for lengthy periods to keep out of the sun as much as possible. Tanning is quick at sea due to the salt atmosphere and light reflection, so there is little need to sunbathe. The other point to remember is that with cooling breezes the sun appears to loose much of its strength. It's a trick of course, the sun is burning your skin and doing considerable harm – the only difference is that for the moment you can't feel it! It is sensible to carry on board a selection of sun creams in the high factor numbers, plus some total block for high risk areas, such as the face, nose, arms and hands.

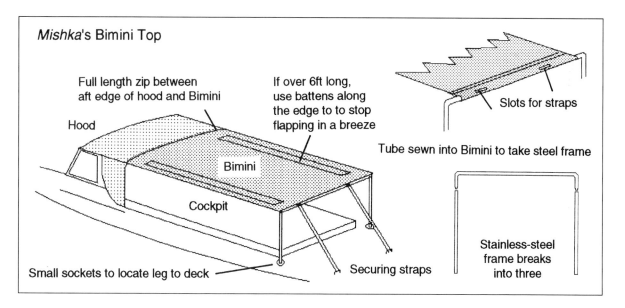

Mishka's Bimini Top

Full length zip between aft edge of hood and Bimini

Hood

If over 6ft long, use battens along the edge to to stop flapping in a breeze

Slots for straps

Tube sewn into Bimini to take steel frame

Bimini

Cockpit

Small sockets to locate leg to deck

Securing straps

Stainless-steel frame breaks into three

We have developed an excellent demountable Bimini top that only takes a few minutes to erect. It utilizes the existing pram-hood with a three piece stainless steel tube as a frame at the aft end, located by small timber sockets screwed to the deck. Tension is by two straps from the aft end of the Bimini to the pushpit rail.

We put a great deal of thought into the design of our Bimini. The main priorities were to make it simple to erect and as quick to dismantle as possible. We also wanted it to be left up whilst sailing. As our yacht has a mizzen mast, it made the design more complicated, as we have to get to the mast and sails when the Bimini is erected to tend the halyards etc. The original thought was to have one permanently mounted on hinged stays that stow flat on the afterdeck when not in use. This design was ultimately rejected, as we felt it would get in the way. Finally a design was agreed upon that utilized the existing hatch dodger as a canopy. We had a full length zip put on its after edge to which we attached a length of canvas to fully cover the cockpit. This in turn was supported at its after end by a dismantleable stainless steel hoop (in three pieces) which located into small deck sockets to stop it falling forward. Twin Dacron stays were then run aft to the pushpit top rail. The whole contraption takes just a few minutes to put up and take down, but has proved to be a brilliant success. It has remained on board and erected, in winds of forty knots or more; but we usually bring it down for safety sake when the wind pipes up to around Force 7.

An amendment in the mark one design was to fit some small tapes under the hood and roll up the front face. This gave us ventilation from wind blowing from the bow. A mark two version is in the design stage and we propose to do away with the restraining straps on the pushpit and keep the canopy ridged by introducing two stainless steel braces along the outer edges. No doubt by the time this book is published we shall be into the mark three version or further! The concept however does work and I commend it to anyone who cares to take the time and trouble to have it made up. Certainly single masted yachts, particularly ones with aft cockpits can put the thing together quite simply.

Awnings and Materials

Overall awnings can be very useful to keep the yacht cool whilst anchored, but if not correctly designed can be rather time consuming to put up and remove. The first requirement is light weight in the material chosen. This will also make it easier to store when not in use. Don't choose polyester sail cloth materials, they are very noisy in a wind and will drive the crew demented. We have a wind scoop made of this material and although efficient , it can make sleeping below a hit and miss affair due to its constant crackling and rustling noises.

When designing an awning it is best not to make each section too large or it will be so difficult to erect that after the first time or two it will not be used again. On a thirty five foot yacht or over make it in two sections with a simple lacing to join the sections together. The shape is not too important other than to ensure it does not overlap the width of the boat. Resist the temptation to have built-in side curtains as these will hold in the heat and keep the cooling breezes out. Side curtains are useful for keeping the low evening sun off, but they can be made as separate pieces that can be attached to the main awnings by lashings or velcro strips. On most boats some form of athwartships stiffening will be required to keep the forward facing ends of the awning stiff. If this is not done the whole cover will start flapping and vibrating in a breeze, causing so much noise and annoyance that it will have to be derigged to restore some peace to the ship.

We have found three-foot aluminium poles jointed in the middle, with holes drilled at the ends for guys to be attached, works very well. These poles are simply slid into pockets sewn in the awning at both ends and the middle. Make sure all eyelets for the lashings are well reinforced, for the awning will come under a lot of strain when the wind blows hard. A useful idea for long term cruising yachts, is to sew a small funnel into

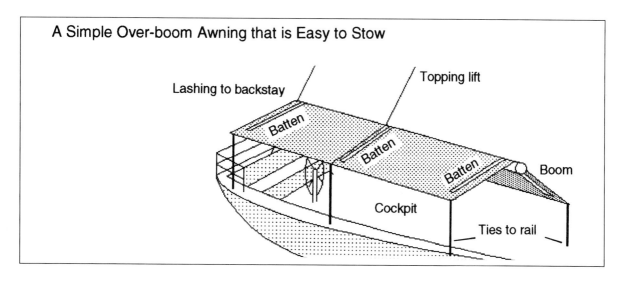

A Simple Over-boom Awning that is Easy to Stow

Lashing to backstay

Topping lift

Batten

Batten

Batten

Boom

Cockpit

Ties to rail

A good awning is one that is simple and quick to erect or remove, and keeps the sun off the cockpit and hatch areas. A full-length awning is awkward to put up and restricts movement on deck; it is only really worthwhile if the boat is immobile for any length of time.

the centreline of the awning for catching rainwater. When it rains, simply slack off the guys holding the sides of the awning a little to make a centre gutter, then push a short length of plastic pipe into the funnel end, and lead it to the water tank inlet, and 'hey presto' fresh water for topping up the tanks. This idea is particularly useful for the Caribbean where watering holes can be few and far between and where short sharp showers are frequent.

Refrigeration – Keeping the Food Cool

Keeping food and drinks cool becomes a serious priority the further south one sails. This not only applies to stopping fresh food from deteriorating, but also to keeping the crew happy with adequate supplies of cold food and drinks.

Unfortunately many yachts have ice boxes or refrigeration systems that have been designed principally for northern climates, with the result that their insulation is wholly inadequate. This may go unnoticed by the average yachtsman or yachtswoman who perhaps uses the refrigeration for a few

weeks in high summer. But the situation is far more crucial when the weather is hot over a long period of time.

In the tropics and Mediterranean, the sea will normally reach temperatures of 75°F or 80°F during the season, and stay there day and night, this means a yacht will be sitting in what is in reality a warm bath. Add to this a daytime temperature of perhaps 100°F and even 120°F inside the boat and you begin to see the problem facing the cold box to stay cool. It is worth looking at the yachts refrigerator or ice box well before the trip begins to judge its ability to remain effective in the conditions it is likely to have to work under.

Insulation

It is most important to examine the thickness and condition of the insulation surrounding the box. In my view six inches of efficient insulation is the minimum requirement, plus a well fitting lid for top opening units. If the insulation is in poor condition consider having it replaced with sheet polyurethane foam. This must be at least four inches thick, six if space is available, and when fitting the foam make sure all the joints are a tight fit to ensure there is no leakage. Do not stick the foam directly onto the inside of the hull, for it will become warm from the sea outside. Make an insulation gap of an inch or so, to ensure a good flow of air. This will also reduce possible condensation problems as well.

It is sensible to pay attention to the top of the compartment for this is where the greatest air loss occurs. The box lid should be a very close fit and be capable of being firmly locked down. If the design of the compartment makes it too awkward to use block or sheet foam efficiently, there is an excellent two-part pouring polyurathane foam available. This is a wonderful material and when the two components are mixed together can be poured into a cavity. A chemical reaction takes place and the foam quickly expands to fill every nook and crevice. It is ideal for areas that are irregularly shaped. A word of warning however, the chemical reaction takes place quickly so have everything ready before mixing the two liquids. The other point to remember is that the compartment to be filled should be completely sealed up, other than one or two small eighth-inch holes in the bottom, to ensure the cavity is filled without voids. This is a 'secretive' material that will quickly find its way through any major cracks, and you will find much of it will end up in the bilges; where it will be the very devil to remove.

Refrigeration Design

If your boat is fitted with a front opening refrigerator, consider carefully the advantages of changing it for a unit that opens from the top. These front opening units were originally designed for caravans where power consumption is not an issue. In a boat, electricity is usually in short supply and every skipper becomes an expert on the conservation of power. These front opening fridges, although convenient to use, are power hungry. This is not a problem when hooked up to shore power, but offshore they can be hopelessly uneconomical. In addition, every time the large door is opened a considerable quantity of expensive cold air flows out, which has to be replaced with valuable amps from the batteries. Some recent models have an additional inner seal which helps conserve the cold air.

Ice boxes can be successfully converted into refrigerators by installing a small 12 volt electric compressor unit. These can be fitted into a convenient locker that is adjacent to the box in question. As these units will generate some heat, the locker should be well ventilated. Expect these units to run for thirty minutes in every hour in hot climates and calculate the amps used when working out battery capacities. Most companies seem to publish rather optimistic power consumption figures for the equipment, so it is wise to bear this in mind.

Other methods of producing refrigeration are from gas or paraffin (kerosene to our American cousins). These require a constant pilot flame which to my mind is a great disadvantage on a boat. It is true that gas models are usually fitted with a fail-safe automatic switch, but my doubts remain. Another problem with these units is that they generate a great deal of heat, which in hot weather is not what is needed. A further method which is really only suitable for larger boats is a mechanical refrigeration system belted direct from the engine or generator; the energy is then stored in holding plates or separate batteries. Whatever system is finally chosen, make sure the capacity of the box is matched to the abilities of the power supply. If in any doubt call in an expert to calculate the sums.

Ice Boxes

Ice boxes can be efficient if adequately

insulated, and make sense where supplies of ice are readily available. The Caribbean is now becoming an area where ice can be purchased relatively easily, the Mediterranean rather less so. One way to reduce 'melt down' is to pack a plastic bag with about two inches of ice and place this on top of the contents of the box under the lid, this helps insulation, but of course it will need replacing every couple of days.

A way to keep the refrigerator cool without expending too much power is to fill the lower section with about six inches of ice and place the contents on top in the usual way. Then run the power unit for a few hours each day. This seems to work quite well and helps reduce power consumption.

Whatever method is adopted, some sort of cold box is a necessity on board a boat in summer. A cool drink, be it beer, water or whatever, is a wonderful refresher that raises the spirits; it makes a contribution towards enjoyment of on-board living out of all proportion to its cost and inconvenience.

Chapter 11
What To Do Before You Leave

Rigging, Lights, Spares to Take, Maintenance, Paperwork, Money, Insurance, Guns, The Crew, Watches & Watchkeeping.

The overall objective is to make yourself as independent and self sufficient as possible.

The boat will be your home for some time, so ensure it is as comfortable and safe as you can make it. In particular don't underestimate the time it will take to make all the preparations and alterations. Work to a list, calculate the time it will take to complete all the work, then double it. Unless you are in the happy position of being able to give the project full time attention, allow at least six months to prepare the boat, plan the cruise and organize the crew. To leave in June, ideally you should start working on the plan from the previous October.

Getting the Boat Ready
The average boat in good seaworthy condition usually will need some extra work to make her ready for an open water cruise to the Mediterranean or Caribbean.

Firstly, prepare lists of jobs to be done on board. Start with the ones that will require extra equipment or gear, these can then be ordered well in advance to avoid hold-ups later on. Look at your basic equipment with a very critical eye, anything that is suspect, worn or unreliable should be repaired or replaced. It is infinitely easier to obtain spares and have repairs done before setting out than to obtain them abroad.

Check over all the rigging thoroughly, there are many reliable rigging firms that will give expert advise, so if at all concerned about the rig, obtain an experts opinion. Wire up, and tape all bottle screws and rigging shackles.

The same principle applies to the engine. At the very least it should be serviced and the propeller given a visual check for any nicks or damage. Carefully go over the gas installation and check for leaks or wear in

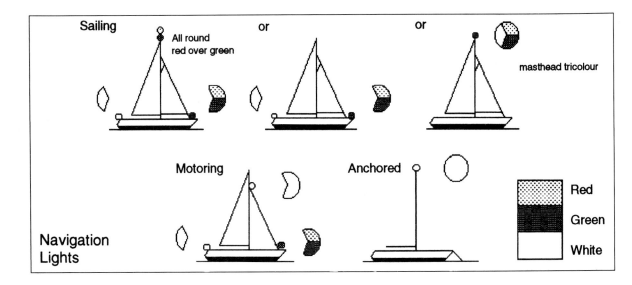

Every yacht venturing offshore should have effective, properly installed navigation lights.

the supply hoses. Send the sails to be cleaned and repaired as necessary. If working sails are ageing, consider replacement with new or get the old sails seams triple stitched. Clean the heads and service the units, look carefully at the inlet and outlet seacocks, service them with all the other seacocks on board, and ensure all hoses are double clipped with high tension stainless steel jubilee clips. The boat should have at least two reliable bilge pumps. Remove as much loose debris as possible from the bilges, then paint them with a good quality bilge paint such as Danboline™.

Now is the time to fit those extra lockers and shelves you have been meaning to make for years, for you are going to need every one.

Navigation lights and wiring should also be examined, and if a powerful searchlight is not in the ships inventory, get one for the voyage ahead. If the yacht has wheel steering, carefully examine the wire runs and remove any backlash by tightening up the system, there are usually adjustable points and bottle screws on the rudder head quadrant. Grease the sheaves, and make

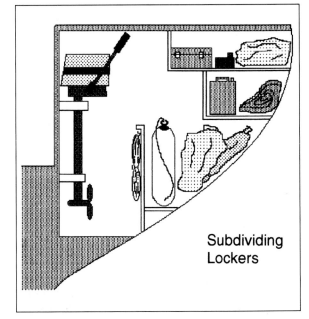

Deep cavernous lockers are not a good idea on a boat as the gear gets dumped on top of each other so nothing can be found quickly. It is a relatively straightforward job to subdivide lockers into separate bins for outboard motors, fuel cans, sails etc. Shelves can be fixed on runners to slide the length of the locker to maximize storage.

sure there are no broken wires, if there are, replace the complete system, just to be on the safe side. Check the rudder bearings for wear, also the prop shaft bearings by firmly grasping the propeller in both hands and moving the shaft up and down. If there is any appreciable movement, probably the cutless bearings will have to be replaced. If the self steering unit is more than five years of age, send it back to the manufacturers for a service.

Carefully check over the bottom of the boat for signs of decay if built of wood, check for corrosion on steel or aluminium, and for osmosis on GRP. Now, not later, is the time to take remedial measures. If all is well, apply at least two good coats of top quality anti-foul paint, and at the same time replace the anodes. It is also a good plan at this point to think about the conditions you will be enjoying once the voyage is completed. A fresh water shower is well worth fitting to the existing cold water supply for showering after a swim. An alternative is to buy a simple shower unit that can be hung on the rigging. It works by filling up a black plastic bag with water, letting the sun warm it up, then using the attached hosepipe that is fitted with a rose, to shower with. Measure up and order any awnings you will need. Service the existing pram hood and cockpit dodgers. Have the life-raft serviced. Replace any out-of-date flares and build up stock to a suitable level for offshore cruising. Service any dinghies, outboard motors and generators you will be taking. Some form of mast climbing steps are really useful to enable the crew to look after the rigging and mast under all conditions. We use a set of Trackstack™ mast steps that have proved very secure in use. Roller furling systems don't seem to need much in the way of maintenance; but look them over carefully to see that all is well and there are no obvious cracks or loose nuts, pins or shackles.

Spares to Take

The ideal situation would be to have a spare for virtually everything that can possible break down or wear out. This of course is ridiculous, but it does make very good sense to cater for the most obvious cases 'of potential breakdown.

A sensible spares list might contain:

The Engine: Water pumps, impellers, fan belts, a complete set of cylinder-head gaskets, fuel and oil filters, spare can of oil for gear box and engine. Injector, thermostat and comprehensive engine manual (not the standard owners handbook).

The Rig: Spare rigging screw for each size fitted to the rig. Bronze seizing wire. Length of rigging wire to maximum length of any stay on board, with spliced eye at one end. Half a dozen bulldog cable clamps. Splitpins in several sizes. A pair of emergency cutters, the best are the parrot nosed ones that can cut through wire as well as stainless steel. Spare halyard to maximum length of the rig. A spare set of Norseman or Sta-lock terminals.

Sails: Full repair kit, some spare canvas plus rolls of tape. Spare mainsail and mizzen slides and one or more battens. Anti-chafe material (we use old towels).

Below Decks: Complete set of washers, gaskets etc. for every head on board. If a pressure water system is fitted, a complete spare pump should be carried, plus an extra pressure switch, motor and gaskets. This sounds extreme but these pumps are put to a lot of use and spare parts are almost impossible to obtain abroad. We have been caught out in this way and it caused a lot of trouble on board without the convenience of having water pumped direct to the various basins. A manual pump back-up should also be fitted, so in the event of electrical failure, at least the water in the tanks can be got at, without having to bale it from the tanks with a cup or bucket.

All internal light bulbs should have spares, plus of course batteries for torches and other battery powered instruments.

Refrigeration will become increasingly important as the weather begins to heat up; so if your boat is equipped with a refrigerator or freezer, contact the agent or makers and discuss with them the spares you should have for your particular model.

Cookers are generally pretty reliable but if yours is elderly, again, contact the makers and discuss spares with them, as you can be sure, if it does break down, no one will have anything remotely like the vital part you are needing so urgently.

General

A good range of galley equipment is necessary including some kitchen scissors (often forgotten until too late). Take some spare cutlery and crockery, we use plastic plates and bowls for eating in the cockpit and china for below decks, this way it keeps the breakages under control.

A comprehensive tool kit should be taken, ideally housed in its own compartment. This must include spanners to fit every nut on the engine, plus two pairs of adjustable spanners, two mole grips, three pairs of pliers, one a long nose type. A range of screwdrivers, slot and phillips, plus some very small ones for electrical work. Small portable vice, a sharp chisel or two, two hammers, a small handsaw, one large and one small hack saw with lots of spare blades. A hand blow lamp, set of Allen keys, tape measure, wire brush and also hand-drill with selection of bits. Every skipper will have his favourites, but tools are essential items that should not be skimped on when fitting out.

Take on board a selection of bolts, nuts, washers and screws, also a range of shackles in various sizes. Bronze siezing wire and some light galvanized wire can be very useful for strapping booms etc. Put some electrical connectors in the bosuns lockers and a pack of epoxy putty that will

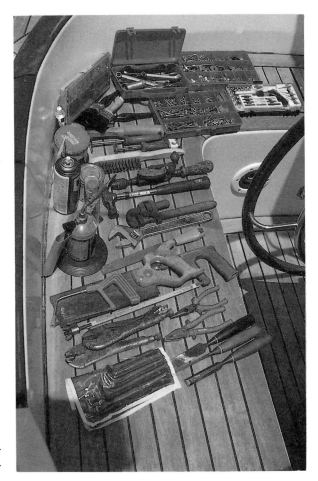

A comprehensive collection of tools to enable the owner to handle most simple repairs on board is vital.

set under water. Also a spare winch handle or two and most importantly, an extra bilge pump handle.

We have some short lengths of plastic hose of various diameters tucked away. They come in useful in all manners of ways, particularly for anti-chafe guards on mooring lines. Include a spare set of genoa sheets plus a selection of warps in good condition. Have at least one nylon warp of one hundred and twenty feet length on board. If there is room, include a fathom or so of chain, it can be stored in the bilges and will come in useful if some semi-permanent moorings need to be made up.

Anchor windlasses have a hard life exposed to the elements at the 'wet end', and they usually receive the minimum of maintenance, so if there is time before departure remove the whole unit from the boat and take it home for a complete overhaul. Contact the makers and get a spare kit of bearings, they cost very little, but will transform an elderly windlass and reduce the power, electrical or manual, necessary to make it function properly.

Grease is wonderful stuff so have a full can on board. Genoa winches need a specialised type that can be obtained from chandlers, it is thin and will continue to lubricate the bearings even under high load, and being very sticky will not wash off like ordinary auto grease. For some reason the aerosol type sprays like WD40 are difficult to come by in certain areas of the Mediterranean, so take an extra can. For some years now we have lubricated our main and mizzen slides with a silicon spray, it is wonderful stuff and has the additional benefit of being non staining or greasy, we purchased ours from a local sailmaker, but I am assured it is also available from chandlers. A flat hosepipe on a cassette reel is also a good idea, but don't forget to include a number of different hose connections, as no two marinas seem to be alike concerning their tap connections.

The list can continue almost endlessly but to be sure all the relevant spares for your boat are carried, sit down quietly on board and mentally go over all the equipment that is fitted, then ask yourself if anything extra will be needed.

Fresh Water

Having enough fresh water storage is sometimes a problem on yachts as most boats are built with coastal cruising in mind, where a water tap is usually just a day's sail away.

It is surprising just how much water one person will use in a day, particularly if pressure systems are built into the boat. One way of reducing the overall consumption is to plumb in a salt-water pump, this can then be used for cooking and washing up. (To avoid long term corrosion problems with the cutlery and pots always rinse with a little fresh water).

Restricting the number of showers to only one or two a week when on passage, will also save on water, although this decision has to be offset against the crew starting to become slightly 'ripe'.

Tankage should be split up into more than a single unit. It's not at all unusual for the contents of the water tank to end up in the bilges when cruising offshore. We always carry a couple of large cans of fresh water as an emergency supply, just in case. On average, we assume a minimum consumption of one gallon of fresh water per day, per person as the minimum requirement, and work out our tankage from there.

Keeping the Water Fresh and Pure

Some areas in the Mediterranean have rather poor water that is not fit for drinking without further purification. This can be done in a number of ways, but perhaps the simplest is to add chemicals to the tanks as they are filled. Aquatabs™ are good and generally easy to obtain. The other more permanent method is to fit a dedicated water filter system that will screen out impurities. There are a number on the market but care should be taken to select a system that has been designed for use on yachts, and if of the in line cartridge type, replace the cartridges annually or to the makers instructions. This is particularly important if the cartridge is of the activated carbon type and does not contain silver.

It is a good plan to thoroughly flush through the whole water system before leaving with a cleansing agent such as Puriclean™. This will kill off bacteria and make the pipes and tank ready for treated, clean water again. We tend to use our tank water for washing purposes only, then for drinking, buy bottled water in convenient one and a half or five litre plastic bottles. They are very easy to store and can be obtained virtually anywhere in the Mediterranean.

Equipping Yourself

The paper work for the trip should also be considered early. There is not a great deal of it but a start should be made about three months before the date of departure; just in case there are any problems that need time to deal with.

All the crew should have current passports, up to date that is for their return, not only for departure. For UK passport holders, no visas are required for any country in the Mediterranean apart from Libya, Egypt, Syria and Albania. For US passport holders Algeria should be added to the list. These regulations alter from time to time and are dependent on the current arrangements agreed between the countries, so it wise to check at the Home Office or equivalent well before departure date.

Customs should be notified in the UK or home country and clearance forms completed. This is normally pretty straight-forward, but is necessary to avoid any problems when the boat is brought back to home waters.

Some form of boat handling competence certificate is very useful. Any of the official Yachtmaster qualifications will be fine or the RYA Helmsman's Certificate will do. They are not asked for in the majority of countries but just occasionally some official in an out of the way port may ask for it, so it's wise to be prepared. Petty Officialism can be rife and red tape is one of the burdens of cruising abroad, but don't let it get you down. We have found that a little patience, plus plenty of smiles and good humour will get you through the local procedures faster and more efficiently than a show if impatience and ill temper. A good tip is to have an attractive wife or daughter with you when clearing papers, it's quite amazing how pleasant even the most awkward and grumpy official can become when faced with a dazzling smile from the fairer sex! The thing to remember is that you can't buck the system. If six copies of all your papers are needed, then six copies it will be, no matter how futile you may think the whole procedure is.

Ship's Papers

Some official document detailing the ships name, home port, registration number and proof of ownership is required in all countries. The standard Lloyds Blue Book is ideal, or papers from the Small Ships Register which is administrated by the Driver & Vehicle Licensing Agency (DVLA), should prove adequate. A yacht's name stamp or headed paper is quite useful for it helps to keep officialism at bay, particularly in the more red tape ridden countries of the eastern Mediterranean. Official looking papers work wonders and I am a great believer in playing them at their own game. The bureaucratic web of rather underpaid local officials is widespread so we always try to understand their problems and provide the papers they need as efficiently as we can.

Insurance

In the Mediterranean, full insurance and an up to date certificate in the skipper's name will be required virtually everywhere. Check with your insurance company on the exact cruising limits that apply to your cover, for the further east you go the higher the rates! For ocean crossings, which include the trip from northern Europe to the Mediterranean, the normal crew requirement is a minimum of three experienced persons.

If you lend your yacht to friends or family, give them full written authorisation in the form of an official letter, and name one person as skipper for an agreed period of time. This is particularly important if the yacht will be cruising to different countries during their period of occupancy.

Money

If your plans are for long term cruising that will keep you away from your home country for lengthy periods, then some form of financial arrangements should be made with your bank to make funds available abroad.

If spending a few weeks or perhaps a month or two on board cruising, then the most convenient method of arranging cash is via Eurocheques and a Eurocheque card. Most towns in the western Mediterranean have changing facilities and many in the eastern Mediterranean too.

Charge cards such as Visa, American Express etc are accepted widely for goods, but for cash they are less useful. We have a supply of travellers cheques on board; the best are in Sterling, Dollars or Marks and in some places Swiss francs are readily acceptable. Cash in hard currency is of course always acceptable and an emergency supply should always be kept on board for purchases in out of the way places. I can't see the owner of a small beach taverna in Turkey or Greece in some out of the way anchorage being particularly thrilled at the sight of a credit card being offered for payment of the evening meal!

Guns

In my view firearms have no place on a yacht cruising in the Mediterranean or for that matter the Caribbean. I know there are some instances quoted of piracy on the high seas where drug running gangs take over private yachts and kill the crew, but these are extremely rare and occur in high risk out of the way places. A yacht cruising in the well populated traditional cruising areas should not have any trouble.

The pro-gun lobby site instances where producing a shotgun would scare off potential thieves, but in my estimation it would, in all probability, cause more of a problem than it is likely to solve. Can you imagine the average drug runner being scared of a shotgun being waved in their direction? The chances are they would consider it a declaration of intent and shoot up the yacht with their machine guns with disastrous consequences for all on board.

Most yachts carry some sort of flare launcher or Very pistol which might be of use in a real emergency, but even here I would be very cautious in showing it off. In some countries the authorities take an extremely dim view of firearms on board yachts and will impound them when the yacht enters their territorial waters. Some of the politically sensitive countries of the Mediterranean such as Libya, Israel, Algeria and so on take a very hard line and it's just asking for trouble to have any form of firearm on board.

It is quite usual to be asked by officials when entering their country if any guns are on board the yacht. If there are any, they must be declared, for if hidden and subsequently found in a customs search, the skipper and possibly the crew could face a prison sentence, which apart from anything else would ruin the cruise!

The Crew
Watches and Watchkeeping
The crew selection of course is a vital

ingredient in making an extended cruise a success.

My experience both racing and cruising points to compatibility as one of the main ingredients for a successful cruise. It's all very well having members with considerable sea time, but if they don't 'gel' with the other members, the ship will never be a happy one and probably not very efficient either.

Some people are natural raconteurs, able to tell a good story and always enjoying the sound of their own voices. Fine for a weekend with the boys, but very wearing after a week or so at sea. Others find great pleasure in winding people up, a tendency that must be restrained at all costs in the confines of a small boat if harmony is to be maintained.

It's no exaggeration to say that very many long distance cruises end up with serious discord among the crew. We all start off great friends and with the very best intentions, but as the days go by, little irritations that would be ignored on shore build up out of all proportion, and cause arguments and unhappiness that ruins the enjoyment of others.

I well remember a particular rough North Sea race some years ago. I was watch captain on a large, heavy Class One yacht, crewing for a skipper who enjoyed a reputation of being somewhat tough and insensitive. We had not sailed together before and I got off to a rather bad start with him, for I wanted to get all the crew together before the race for a briefing on where all the gear was and how it all worked. This was not his style at all; he believed we should have known this (although I am not quite clear how) before the start.

As the race progressed the crew began to succumb one by one to seasickness, until there were only three of us on our feet out of the original muster of nine. The skipper, who felt not a twinge of queasiness himself failed to understand that others were ill (in fact one poor chap collapsed and was taken

to hospital to recover when the race was over) and accused them of malingering. He insisted they attempt to stand watches in the normal manner, and it was only when he retired below for a rest, did I manage to restore some order and get the sufferers to their bunks.

The weather deteriorated further, as did morale and the boat's performance. The irate skipper continued to behave badly and to berate the crew. At one stage I thought my next duty would be to quell a potential mutiny, so badly had the situation deteriorated.

Of course it was no great problem as it was just a two day race, we just had a miserable weekend at the hands of a self indulgent and very poor skipper. But had we been involved in a long race or extended cruise goodness knows what would have been the outcome.

This of course was a rather extreme instance, but does illustrate what can happen if a skipper, or for that matter an individual crew member is incompatible, and through insensitive behaviour completely ruins enjoyment for the other members. I suppose the old axiom 'Think of others before yourself' is a truism that is particularly pertinent in the confines of a small yacht at sea.

Watches and Watch Keeping

Watch pairing is a matter every skipper has his own ideas about. If the crew varies widely in ages it is always a temptation to pair off the watch members with contemporaries. This can be a mistake however, particularly if you have reason to believe the trip will involve hard weather. I try to balance out every watch couple with experience and physical strength. If for instance you have two middle age crew and two youngsters, much better to give each watch a strong youngster for those difficult times when a little extra agility and muscle can make all the difference. This also applies to experience. Always try to keep one experienced person on every watch whatever the system employed, particularly with night sailing. Judgement in difficult situations, and there always seem to be a few, even in the best run ships, is often helped if one crew watch member has been in a similar situation before.

Watches of course will depend on what crew is available, the length of the cruise, and their individual abilities. I am a great believer in keeping the skipper, particularly if he is also the navigator, out of the watch system completely. This will mean he is well rested and able to make rational decisions that are not clouded by fatigue. He or she will also be ready to relieve other members of the crew at odd times to give them an extra hour or so rest, also help prepare meals and attend to the general maintenance and running of the ship. All extremely valuable duties that will be appreciated by the normal watch keepers.

I have illustrated some watch keeping systems based on methods that have been used successfully in the past. Every skipper will have his or her preferences but I feel the following points should be considered carefully.

Once the crew has been agreed and the watch system chosen they should all meet before the cruise starts to be briefed by the skipper:

1. A copy of the watch system, complete with times etc. should be given to all members.

2. Every member given a specific duty if an emergency such as abandoning ship arises. i.e. one person gets the life-raft over the side and inflated. Another member collects all the life-jackets and issues them out. Another picks up the emergency panic bag and EPIRB. Yet another person makes a Mayday call on the radio telephone. In this way there is no panic, everyone knows what is required of them and hopefully no one is injured rushing about. Of course overlaying all this is the instruction that it all takes place only if ordered by the skipper and no on else, unless of course he or she is incapacitated.

3. The crew is carefully taken over the ship and shown how all the equipment works (assuming of course that they are new to the yacht).

4. If some of the crew are unfamiliar with R/T work they are given instruction.

5. The route plan is thoroughly discussed with all members not only to get their ideas, but also to reassure them that their involvement in the planning is welcomed just as much as their muscle.

6. A full discussion on meals they are likely to get dished up, food to avoid, how many meals during the day, individual preferences and the role and consumption of alcohol.

Each skipper will have his preferences, so this list will be extended. The main priority is to give each member of the crew a feeling that their contribution is important to the team effort. It also has the very real advantage that each crew is fully conversant with the running of the boat. This removes a great deal of anxiety from the skippers mind. The poor chap will have enough on his hands making the cruise a success without concerning himself about the competence of each individual crew member.

Chapter 12

Living On Board

Sleeping, Heads, The Galley, Stowage, Gas, Lighting and Ventilation

The most important aspect of living on board for extended periods at sea, is for the crew to be able to rest properly, prepare meals easily and eat comfortably.

Offshore sailing can be an exhausting pastime, even in moderate weather, due to the incessant movement of the boat and the watch system which disturbs the normal cycle of sleep. Obviously with a large more stable boat, things improve dramatically, but for most of us we have to make the best of things with the boat we have.

Speaking personally, I would not enjoy the prospect of going offshore for an extended passage in a yacht smaller that thirty two feet, although many hardier persons than myself do so regularly and enjoy the experience. It is not a question of seaworthiness, but more one of comfort. Perhaps there is some truth in the old saying from the Victorian era of yachting, that 'For windward sailing one needs a foot waterline for every year of your age'.

So what can we do to make our own yachts as comfortable and convenient to live in as possible?

Sleeping

It is essential that every crew member has his own bunk, and that it is of adequate length. Six foot six is about right. It should be sited in the main cabin or aft section of the yacht. No bunk forward of the mast in a small yacht is habitable for any length of time, for the motion will be too violent when going to windward.

In addition, all the berths must be fitted with lee cloths that are secured with strong adjustable lanyards. A good method of attaching lee cloths, is to screw a batten along the cloth on its lower end, tucked well under the berth cushion, then have the lanyards led slightly inboard to curve the lee cloth around the body. This will give maximum support and enables the crew member to lie *into* the lee cloth for relaxing sleep. Lee cloths should not be too long about four feet and about ten inches deep. The head section should be cut back a little to allow for good air circulation. There is nothing worse than sleeping in a cocoon of stale air; you wake up weary and fuzzy and not at all refreshed.

Quarter berths and the old style pilot berths are really the best for sleeping in, being cosy and well out of the way of the normal comings and goings on the boat. I like aft cabins if they have single berths, as they tend to be quiet (if the engine is not running) and in the part of the yacht with the least motion. We are fortunate on *Mishka* to have two twin cabins aft, and they have proved to be very successful. If the aft cabin has a double bunk, it will prove to be useless for offshore work for it is too large and offers little support for the sleeper. Better to take the mattress to someone who specialises in boat upholstery and have it cut down the middle, then fit lee cloths, which will turn it into at least one and possibly two reasonable sea berths. I am not in particular favour of the traditional pipe cot seen in some boats. These are usually canvas areas supported by aluminium poles, and even if they can be adjusted to the curve of the body are devilishly uncomfortable for any length of time. Fine perhaps for the odd weekend but not really

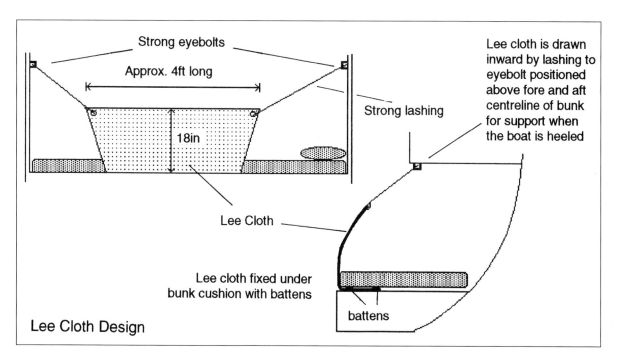

Strong eyebolts

Approx. 4ft long

Strong lashing

Lee cloth is drawn inward by lashing to eyebolt positioned above fore and aft centreline of bunk for support when the boat is heeled

18in

Lee Cloth

Lee cloth fixed under bunk cushion with battens

battens

Lee Cloth Design

Care is needed when installing lee cloths to ensure they give the maximum support to the sleeper. Always attach the top of the cloths well outboard to give plenty of adjustment for sailing heeled on both tacks. Ensure the lee cloths are tucked at least a third under the cushion and secure with a batten to the bunk bottom.

for extended cruising. It is beholden of the owner to provide comfortable berths for each of his crew when cruising for any length of time. In return he will be rewarded with a well rested and cheerful crew.

Handholds

You can't have enough of these on a boat. They make moving and working below, safe and relaxing. The violent motion in a seaway is exhausting at the best of times and if there are not enough well positioned handholds to assist in a safe passage around the boat, then someone is almost bound to fall and be injured.

Take a careful look below, then walk around just holding onto the available handholds – not balancing in the normal way. This will give some idea of just how

many handholds are needed. Don't forget to examine the heads compartment and sleeping cabins plus of course the galley, companion-way and navigation area. Pillars are excellent as they give support to the whole body, additional single handholds can be fitted quite easily, but ensure they are bolted on, not screwed. Full length handholds running at shoulder length down the cabin sides are the best of all, as they give continuous support right through the boat.

Another small but important point, is to make as many corners in the cabin joinery rounded rather than at right angles. It's surprising haw many bruises one collects in lively weather. If the joinery can't be altered, look at the possibility of temporarily padding some of the most obvious places.

Heads

This is an important area in any yacht. For the greatest comfort in a seaway, an aft head is best, but not always convenient. There is little one can do with the construction of a head if it's already built, but there are some modifications that can be made, that will make it more suitable for offshore work. Make sure the floor is as secure and anti-slip as you can make it. If it incorporates a teak shower grating, remove any varnish that may be on it. Bare teak is excellent for secure footing, even when wet. Make sure that there are convenient hand holds at shoulder and sitting height.

Hopefully the wash basin is not tucked too far under the coachroof sides, but provide a secure, deep soap and toothbrush holder. Use plastic glasses for water. As the head is often the place where the crew don their oilskins, make sure there are no sharp projections around, and if there are try to pad them. If possible try to fit a bolt or loop and eye to hold the loo seat in the upright position. Have a good supply of disinfectant or bleach available to keep the area clean and sweet smelling, and try to get the crew to join a head cleaning roster – otherwise it will be down to the skipper!

The Galley

The main priority with the galley is to make it a safe place to work in, irrespective of weather conditions.

I prefer U shaped galleys as the cook is surrounded with all the work surfaces at an arms reach. This means that if he is strapped in with the bum support, there is no need to unclip every time he needs something away from the cooker.

Safety is most important, as hot flames and boiling water are unpleasant ship-mates. It really goes without saying that the cooker should have efficient gimbals fitted and also a bolt or catch to keep it from swinging, when required in port. To keep the cook from falling onto the cooker a robust crash bar should be securely attached, but ensure it will clear the top edge of the cooker when the unit is at it's maximum 'swing'.

An area that is surprisingly neglected on very many offshore yachts is the galley sole. This is all the more surprising when one considers all the water and cooking spills that inadvertently occur in the normal way of things in a ship's galley. Shiny varnish is lovely to look at, but it belongs to marina living, not in an offshore yacht. Stick lengths of 3M non-slip tape on the sole, or do as we have done on *Mishka*, and screw short teak battens about three quarters of an inch wide into the floor. This has worked well and can be easily removed, just leaving a few small screw holes which can be easily filled at a later date if need be.

Deep, twin sinks are a blessing. They hold all the dirty dishes and cutlery until the crew get round to washing up. They are also useful stowage for hot saucepans if the ship is jumping around. Deep fiddles around the working area are also seamanlike, but make sure the inboard sides are vertical and do not slope. It's worth the time and effort before departure to make some sort of rack or holder for filled cups or mugs. This will avoid spills and enable the cook to prepare more than one drink at a time.

Custom stowage for cutlery and pans is also something that will be appreciated by the cook. It will also save many breakages in the long run. If your food cupboards are hinged from the top, change then round to bottom hinging or you will have the contents falling out every time they are opened. Try to arrange a gash bucket into the galley somehow, then fit it with removable bags or liners. We always have a routine for rubbish disposal. All waste food and biodegradable rubbish including light paper (not cardboard) is dumped overboard. Plastics and tins are washed in sea-water then stored in a large black dustbin sack for disposal when we next go ashore. This way we feed the fishes, don't pollute the sea and

The simple expedient of screwing a few teak battens to the galley sole has turned a potentially lethal ice-rink into a secure and safe place in a seaway.

A good galley arrangement for offshore use. Note the two deep sinks, fresh and salt water taps, filtered water outlet, anti-slip mat and simple holder for filling mugs. The thick chopping board is held on the sink with four adjustable blocks that press on the sink sides. It is a design that has worked well in practice.

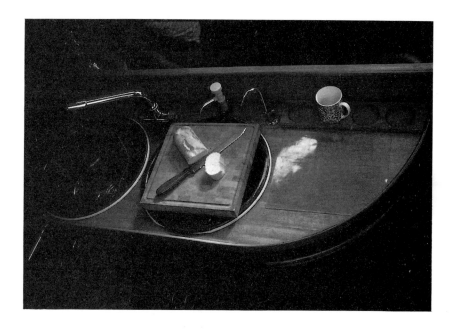

give the recycleable rubbish to the people who can handle it.

Piped hot water is very convenient to have in the galley as it ensures that everything can be kept really hygienic and clean. With engines that are fresh water cooled it is a straightforward matter to connect a calorifier to the system, this then ensures whenever the engine is run for battery charging or whatever, a full tank of piping hot water is available. To keep water consumption down, fitting a salt water pump to the galley is a sensible option. If your particular boat does not have one, it's no great problem to install a unit, but ensure the sea inlet is fitted with a seacock.

Stowage

Food stowage is usually a headache as space is always at a premium; food will therefore have to be stored all over the boat, utilizing any spare lockers that are available. To help overcome the resulting confusion make up a food locker chart, and list where all the spare food is stored with a detailed description of their contents, then pin it up close to the galley.

It helps with the stability of the boat if the heavy items, such as liquids, tins etc are stowed lowdown in the bilges or lockers. If there is a likelihood of the bilges getting wet, and what boat keeps really dry, remove the labels from the tins and mark the contents on the lid with a permanent marker. Tins rust very quickly when in contact with salt water or damp salt air, so be careful to inspect old ones, before consuming the contents. If it is proposed to keep tins on board for longer than six months, they should be protected from the atmosphere in some way. Varnishing is the time honoured method; it's messy but effective.

If the cooker is fuelled by gas, have a remote controlled shut off valve installed close to hand so that there be no need for the nightly trip on deck to turn it off at the gas bottle. Gas bottles should always be installed in their own leak-proof lockers, with a run off outlet piped over the side.

If space allows, make up a permanent stowage for a thermos flask. This can then be filled with hot tea, coffee or soup, wonderful for cold night watches! It can be prepared after the evening meal when everyone is awake, so avoiding waking the off-watch crew with banging and crashing in the galley later in the night.

Lighting

Good lighting at night is important for the efficient running of the ship and the convenience of the off-watch crew.

A swivel light over each bunk to enable the occupant to read or find his clothes etc is a minimum requirement for sleeping cabins.

Low intensity lighting in the main cabin and galley area is something that should also be considered. There are a number of inexpensive units on the market that use very little electricity, but give enough diffused light to enable the crew to move around without disturbing the off-watch members. Some yachts have small low power red lights at foot or knee level, these are excellent, as the red light does not interfere with night vision.

Chart tables should have an adjustable chart light, if possible with a dimmer switch and red filter. It is important for night sailing that the watch on deck is not blinded with strong lights from below at any time, the trick is to have sufficient light for the oncoming watch, yet not so much that they will take more than a minute or two to adjust to the darkness of the cockpit.

Ventilation Below

Ventilation is most important, yet on some boats it is a neglected part of the fitout. For offshore work it is necessary to arrange things so that the boat is ventilated even when all the hatches are closed down. This necessitates the inclusion of watertight vents such as the dorade. These will pass a surprising volume of air. The main areas that are most difficult to ventilate are small aft cabins, the heads and galley. In fine weather the main hatches can be opened and the boat ventilated through, but this is not always possible if the boat is sailing to windward in rough conditions, or if it is raining heavily. Small passive or solar powered vents are most useful in these conditions and can be fitted quite easily. It is important that a flow of air is maintained at all times to keep the boat sweet. Some form of ventilation system should be installed for sailing under wet conditions when all the main hatches have to be closed.

Oilskins get damp, and if possible they should have their own locker for drying out. If no special place has been allocated in the original design of the boat's accommodation, put up some strong hooks in the head compartment and hang the oilskins on plastic hangers. The ideal place of course is by the engine, where its heat will keep them dry and aired, but this is possible on only a relatively few boats. What is important, wherever they are kept, is to ensure the space is well ventilated, otherwise mould will grow both inside and outside of the garments very quickly, particularly in the tropics.

Once you arrive in the sunshine, be it the tropics or Mediterranean, ventilation becomes a big issue. Yachtsmen used to northern European climates are usually concerned about keeping warm. But from now on the priority will be to keep the boat and yourselves cool and fresh. The Mediterranean weather, although hot, is very different to the Caribbean, where the trade winds blow practically continuously which can result in almost air conditioned comfort.

In the Mediterranean the winds are less predictable, and there are many days when there is little or no wind at all. This is when you quickly get your priorities in order, for to be below on a small boat in stifling hot weather with inadequate ventilation can be torture. The need is for hatches, big ones, to let in the maximum amount of air.

Ideally, every cabin and closed compartment should have its own hatch, but sometimes this just is not possible. There is one hatch on the market that is quite brilliant, in that it can be tilted forwards or backwards at will. It is made by Goiot of France and marketed in the UK by Montague Smith Ltd. These are low profile alloy hatches with acrylic tops to let in the light. Their real triumph of design is that no matter if the wind is from forward or aft it can be angled to catch the breeze.

We also use a windscoop to funnel air below; which is most effective. There are as many designs of scoop as owners who use them, and they all have their advocates. We have experimented with several designs and find the commercially available variety works very well. A word of warning however. If contemplating a DIY effort make sure the material used is of a light but soft variety. Some hard spinnaker cloths make a devil of a noise in a wind, then no-one on board will get any sleep!

The important thing is to endeavour to keep the air moving within the boat. A constant movement from forwards to aft will cool a boat down and keep it comfortable. It will also make sleeping below on hot nights possible.

Chapter 13
How To Get There

Northern Europe to the Mediterranean

There are two routes from northern Europe; one through the inland waterways of France, the other, the open sea water route to Gibraltar.

The Inland Water Route

This route via the French canal system is straightforward and is the choice for many motor and sailing yachts who consider the sea route rather too arduous an undertaking. It is on the other hand, quite hard work, for there are many locks to negotiate as one climbs over the mountains en route. Some owners use the sea route out, to take advantage of favourable winds, then the inland route on their return.

Depth of water is a limiting factor and any boat drawing around five feet should check with the French authorities on available water, for the canals do vary in depth from year to year. Air draught (the overall height of the superstructure from the waterline) is another factor, particularly with larger motor yachts with fly bridges or high wheelhouses.

It is necessary to have a reliable engine that can make a cruising speed of around five knots, as the distances are quite large. Locking in and out becomes second nature, but is it hard going if a tight schedule has to be maintained.

At least four good long warps are required as well as six to eight stout fenders. The walls of some locks can be rough so they can give fenders a hard time. Some people use old motor tires but these mark the topsides, unless they are covered with canvas. A gang plank must also be carried, for getting ashore when tied up for the night, and of course, you must have a couple of long boat-hook. Mooring anchors for digging into the banks come in useful on occasions.

There are good craning facilities at both ends of the canal system, so getting the mast down and up again is not a problem. Some secure form of mast support should be constructed before the start, in order that the spars can be lashed down over the coachroof. It is best to have these made up so that the maximum length of the spars are well supported to avoid unnecessary bending and possible damage.

The actual cruising between the locks can be delightful, the countryside is lovely, the waters calm and in general, the lock keepers are friendly and helpful. Provisioning facilities are quite good, but in places they can be quite a distance from the locks. A bicycle is invaluable, for apart from use as a provisioning 'taxi', it can be used to explore some of the lovely areas that adjoin the canal system.

Although less demanding than the sea route, the boat should be in no less seaworthy condition as the entry point may necessitate cruising across the English Channel or North Sea and if the Midi route is chosen, also along the Brittany coast. Before a decision is made to use the inland route it is sensible to read one or more of the specialised books on the subject.

The Open Sea Route

Most yachts set off across the Bay of Biscay from Falmouth, which is a sensible plan, for it is a very good provisioning area and being placed almost on the tip of south-west England is the closest to the Brittany coast. For most yachtsmen the prospect of crossing the Bay of Biscay is not one that is particularly relished, for it has a formidable reputation, that is well earned. If however the crossing is planned during the summer

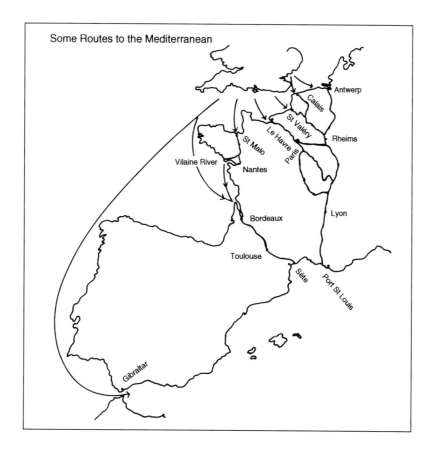

Some Routes to the Mediterranean

months from June through to early September it is not unreasonable to expect moderate conditions. The Bay itself is however a large expanse of water that can be unpleasant, so I personally like to get it over with as quickly as possible, without, unfortunately, visiting some of the lovely Rias and harbours on the Biscay coast (I am saving these for a separate cruise).

If on departure the forecast is for strong south-westerly winds, then I favour taking a long hike on the port tack once the Scillies are abeam, right out towards the Irish coast. This westing can be invaluable later if a south-westerly gale springs up when half way across, as it will avoid the yacht being pushed deeper into the bay, then to face a hard windward slog to round the corner of north-west Spain. If time is not pressing the yacht can rest for a day or two at somewhere

such as La Coruña. This is a large protected harbour with an easy, well lit entrance and good facilities. The main difficulty is in its approach as the shipping lanes converge here and it can be very busy with traffic.

On leaving La Coruña, a decision will have to be made either to close the land for an inshore route, or keep well offshore clear of the shipping lanes. Commercial shipping always take the most direct routes, so they can be relied on to almost keep 'on rails' between landfalls; which means at the turning points on the route, congestion is inevitable. Care must be exercised at these points to comply with the strict rules on separation zones, to disregard them will open the skipper to heavy fines. The run down the Portuguese coast should coincide with northerly winds so that the yacht will be running with boomed out sails most of the way. The coastline varies a great deal, and once the mountainous regions are left behind a low sandy rather uninteresting coast takes over for a while. The big advantage, other than steady winds is that the weather will improve, and it gets progressively warmer the further south the yacht sails.

Finisterre is an area that seems to attract a lot of hard weather so a yacht can expect fresh winds in this region. I favour keeping well offshore, for if faced with a gale at least there is plenty of sea room to ride it out. Vilamoura in southern Portugal is a convenient place to stop over for a rest. It is clean and well run, with good provisioning facilities. I always feel at this point, the weather and landscape takes a Mediterranean atmosphere with warm evenings and lovely days full of sunshine. The run to Gibraltar is straightforward enough, but it can be rough if a Levanter is blowing; in which case shelter should be sought until it blows itself out (usually about two days). Gibraltar is the gateway to the Mediterranean and as such is full of atmosphere and traffic. Facilities are good and some things, notably alcohol are very inexpensive. This is the place to stock up on spirits and beers. Diesel is also reasonably priced but only available for cash; credit cards for some reason are now no longer accepted.

Across the Atlantic to the Caribbean

In many ways Gibraltar also makes an ideal place to start the voyage across the north Atlantic, although Vilamoura in Portugal is used by some yachtsmen. Facilities for repairs and stocking the boat are excellent in both places with Gibraltar being better for some things, such as engineering repairs and chandlery.

The time to arrive in the Caribbean is after the hurricane season, which to be on the safe side is mid-November. This means crossing the Atlantic from say the Canary Islands or Madeira any time after November or perhaps early December. Unfortunately this may prove to be a little early for reliable trade winds to be fully established. However the winter trades can give excellent sailing conditions but may be variable in strength at times. Steadier conditions prevail from January onwards. Although many yachts make this trip each year, it should not be regarded as routine. It is imperative that adequate preparations are made. The boat should be fully seaworthy and the crew seasoned and experienced. Crew fatigue is probably the greatest enemy and it does make good sense to have enough people on board to ensure everyone can get enough rest. On a thirty six to forty five foot boat, four crew members should be a good complement. If all are competent watchkeepers then the boat can be run on the basis three hours on, nine hours off or two hours on six hours off with perhaps watches being run during the hours of darkness using two watch keepers, with one of them resting below. Being flexible in this way, it will ensure everyone has enough time to rest properly and if there is bad weather to face, no one is too exhausted to pull his weight.

Landfall after a long period at sea is always a wonderful feeling, but it does have its dangers. Some Caribbean islands are low lying so don't attempt them at night unless you are completely sure of your position. Also remember that the yacht will be running down wind, so this will be a lee shore. Much better if the approach is at night to heave-to offshore and wait for the dawn. There are currents running right through the islands approximately from east to west, so due attention must be paid to them when close to land. Try to choose one of the higher islands, such as St Lucia or Barbados as your landfall, they should give a good position at least twenty miles off in the normally good visibility of the area.

From the East Coast of the USA to the Caribbean.

In some ways this route, although shorter than the European one, is more difficult for a sailing yacht, due to the prevailing winds.

Many yachts cruise down the eastern seaboard during the summer and make the crossing in the Autumn (Fall). The other route from further north is to sail direct to Bermuda, then down the Atlantic seaboard to the Virgin Islands. The most important aspect of both routes is timing. Leave too late in the season and the weather could be unpleasant in the north. Equally, an early summer departure will risk facing the tropical hurricanes that sometimes sweep up the coast from the Caribbean.

For northern based yachts the other consideration is to make the trip in two separate stages, one to Bermuda early in the season, leave the boat over for a month or two, then sail for the Islands in September or early October. In this way both the hard weather patterns will be avoided.

The best landfall for deep keeled yachts in the Virgins is probably St Thomas, it is well up to windward and has good facilities. The sail can be a hard slog to windward but if timed with care, it should be possible to miss very hard weather. I recommend careful study of the wind and weather maps such as Stanford's Atlantic Pilot Atlas before final route planning is undertaken. Then once on route keep daily updates on all the available weather forecasts.

It is likely that on any offshore passage some hard weather and possible a gale or two will be experienced. On a well found boat this should not present a problem provided the boat and the crew have been well prepared. My tactics are always to plan ahead and try to arrange plenty of sea room. Listen to the weather reports and position the boat in the most advantageous area. Then if caught in hard weather, slow the boat down or stop, until the conditions improve.

Trade Wind Sailing

Most experienced cruising yachtsmen agree that it's worth expending the effort to plan a long distance cruise utilizing the fair winds that blow around the globe. It's no fun at all to have to turn to windward for extended periods. When planning a long distance cruise the best book to start with is *Ocean Passages for the World*, published by the British Admiralty. This gives all the recommended routes. Try to obtain a copy that was printed prior to 1914 as it will deal with sailing routes rather than those later editions which tend to be more steamer orientated.

For a sailing yacht embarking on a long cruise that will involve considerable downwind sailing such as the north transatlantic, it is certainly worth considering a specialized rig and sails. In 1992 *Mishka* will hopefully be sailing across 'the pond' and as it is her first crossing I plan to make some alterations to her rig.

Past experience has persuaded me that the traditional down wind twins are not the answer. Because the booms are fixed either on the mast or forestay, they suffer from inflexibility and are difficult to trim. Usually the staysails have to be cut on the small side so the boat tends to be under canvassed other than in fresh winds. The other main disadvantage is that they do induce rolling, which can be most uncomfortable and exhausting for everyone on board.

It's little wonder that Eric Hiscock, one of the great cruising yachtsmen of all time, was driven to write 'Running down the trades in *Wanderer II* with twins set, we rolled gunwale to gunwale, making the crew very weary'. It is therefore really no great surprise to see most modern cruising yachts making do with their standard mainsail with the genoa boomed out to windward. This however, is still an unbalanced rig and it does make the boat roll. I do believe that some form of twin headed rig is the best

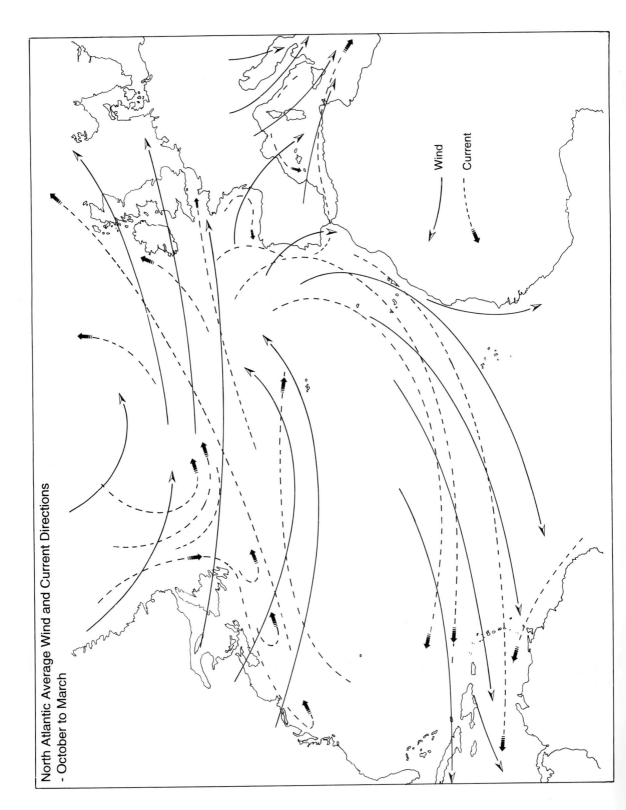

North Atlantic Average Wind and Current Directions
- October to March

Wind
Current

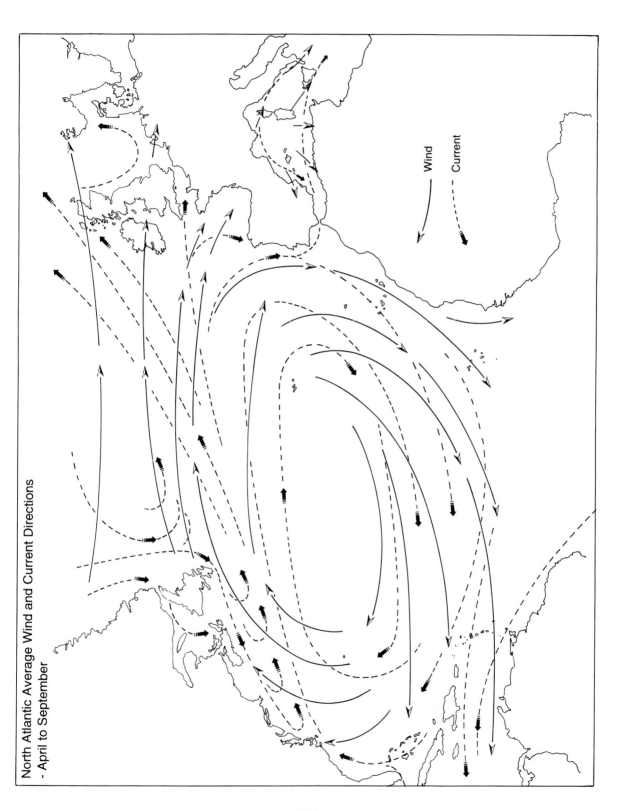

North Atlantic Average Wind and Current Directions
- April to September

Wind

Current

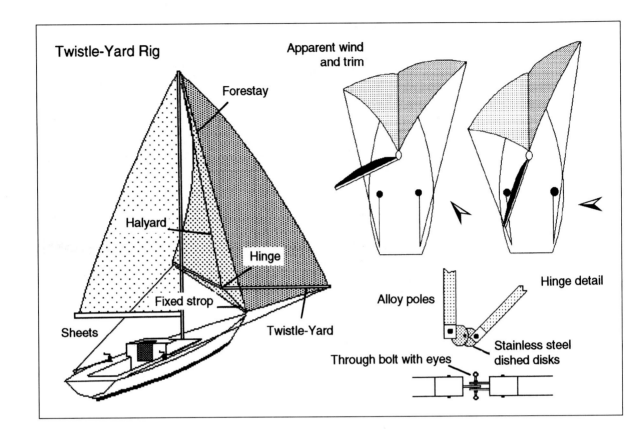

Twistle-Yard Rig

Apparent wind and trim

Forestay

Halyard

Hinge

Fixed strop

Sheets

Twistle-Yard

Hinge detail

Alloy poles

Stainless steel dished disks

Through bolt with eyes

answer, and have been experimenting with a combination of twin poled genoas and mizzen. It makes *Mishka* go, but again is still less than ideal.

On reflection I have decided to adapt the remarkable, but little known, Twistle-Yard Rig (Created by Mr H. Barkla some 20 years ago.) for *Mishka*. It comes with high recommendations so it will be interesting to see if this is the answer for us.

The essence of the Twistle-Yard, is to set two light weather, high cut sails on the forestay, and to control them with a long boom fixed to the sails clews. The unusual part is to come; the boom is not fixed to the mast or forestay in the normal fashion, but is hinged in the middle and set flying, entirely disconnected from mast or hull! The sails have single sheets led back to the cockpit in the usual way, and are hoisted on

genoa halyards. A spare halyard is used as an uphaul for the twistle-yard boom. A fixed vang is than attached under the boom centre and led forward to the stemhead.

This rig is then left to float in the air, and pivot around the forestay. The Twistle-Yard system is so flexible it can be trimmed to run with the wind dead astern to broad reaching, all with the minimum of effort.

It is claimed that the rig is very tolerant of wind shifts, in fact it is so stable, the boat can be sailed round in a circle if need be. One other very important attribute is the ability to dampen out rolling. In hard winds from astern the twins are trimmed to run foreword of the forestay in a deep V, which reduces the accentuated rolling motion that can be so exhausting for the crew. It will be interesting to see how it all works out.

A well thought out navigation area that is practical at sea.
Note the swivel chair that supports the navigator when the yacht
is heeled; large chart table with grab rails that also stop the charts
sliding off; good instrument and book stowage and a large chart
storage locker. All the navigational displays are conveniently to
hand and the bulkheads are hinged for easy access.

Navigation

The world of electronic navigation is moving so fast that any words I write now will be out of date before they are printed, let alone read.

Nowhere in the sailing scene have so many radical changes been seen in such a short span of time. Anyone contemplating a blue-water cruise can be confident that for a relatively modest outlay, a wonderful little black box can be installed that will give the navigator all the positioning information that he is ever likely to need.

But of course this is not the whole story, a prudent, experienced navigator should not rely on his electronic wizardry entirely, for they can shut down, either through failure of some part, the ships batteries can fail or the equipment can get wet or damaged. Then the relaxed atmosphere will change

alarmingly and traditional navigation methods will have to be implemented in a hurry.

Most offshore yachtsmen now equip themselves with Satnav or GPS receivers, for they give quite amazingly good positioning. However the sensible navigator will also continue to plot the ships position on an end of watch basis and cross check any navigation anomalies that occur at every opportunity.

On an ocean crossing there should also be a back up to the electronic equipment, with someone in the crew being competent in astro-navigation using a sextant. If no one has this experience before the departure date, the skipper or navigator should take a course beforehand to acquaint himself with basic techniques, then practise the art as the cruise develops, using the electronic equipment to check against his own astro positions.

There is a great deal of mystery surrounding the black art of astro-navigation, but provided the student has a basic knowledge of addition and subtraction, is willing to put in a little time in learning the basics, then get in some practise with the sextant, anyone can learn.

A good place to start is with a simplified book on the subject, written for yachtsmen, such as Mary Blewitt's 'Celestial Navigation for Yachtsmen', once this has been carefully read and understood, practise with the relevant tables and a sextant, and all will be mastered.

There are some other aids available to budding astro navigators, such as small calculators that reduce the calculation work, and other little machines that have all tables programmed into the memory, so that additional tables become unnecessary.

If your prefer to be taught, rather than learning from books, the well known ocean yachtsman and instructor Geoff Hales will run tailor made training sessions on a one-to-one basis for anyone who can visit his Sussex address. There are also numerous sailing schools that teach the subject, plus postal courses available.

Navigation Equipment

The most important piece of navigation equipment on board is the ships main compass. This should be 'swung' and corrected, as a matter of course before commencing any long offshore cruise.

In the electronics field GPS (Global Positioning System) is a highly accurate world-wide navigational system that has quickly become the favourite of yachtsmen, due to its uncanny accuracy. Originally it was developed for military use but now it is available to civilian users; although I believe it is slightly degraded to limit its usefulness to a potential enemy. The difference with

GPS over the Satnav system is that it relies on a constellation of twenty four GPS satellites in orbit so giving virtual twenty four hour coverage and constant updates. The way it works is that each satellite is available for tracking at any one point for seven hours during its particular orbit and never less than three satellites will be 'visible' at any one time, so giving precise accuracy.

There are three types of receivers available at the time of writing and these get their signals by 'continuous tracking' or 'sequence' or 'multiplexing'. They differ in the way they handle signals and the overall number of receiving channels they have. Few sets use 'sequencing' these days and I consider for the average yachtsman's requirement a 'multiplexing' set will give excellent performance for all normal conditions. 'Continuous tracking' receivers have the reliability and performance standards that are clearly superior to the others, but they are also rather more expensive.

Satnav

This system based on the now decaying TRANSAT satellites which have been wonderfully reliable in the past, although there are plans to keep it in operation until 1996 or later (this is to be confirmed).

The principle is similar to GPS, other than due to the time lapse of the overhead satellites, actual fixing times can be up to an hour or two apart. This disadvantage can be mitigated to some extent by fitting a heading sensor to the receiver that works out elapsed time between the satellite passes, the course and speed of the vessel, and calculates the current position. This is then updated at the next satellite pass. Although it may sound a little confusing, in practice the system works very well and for yachtsmen cruising oceans is a well worth consideration, especially as receivers are now very inexpensive due to the possible limited life of the system.

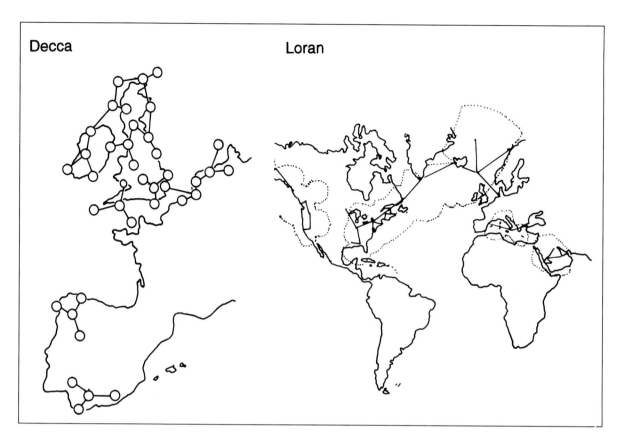

Decca: There is excellent coverage around the UK and English Channel, but in mid Biscay and along the Portuguese coast it can become unreliable.

Loran: In general the Mediterranean is well covered apart from the extreme East, around Gibraltar and western and eastern Turkey where it diminishes. The US eastern seaboard is well covered, but not Caribbean waters.

Decca

If your current yacht has Decca fitted and your destination is the Mediterranean, then this system will get you well down the Portuguese coast. We found it started to become unreliable about mid-way, but then perked itself up again a little further south. The system has been given a new lease of life with an extension to its programme into the late '90s.

Loran

This system has fairly good coverage in the Mediterranean but can be unreliable at the extreme edges of its range. The US eastern seaboard is well covered and it has been the principal navigation system there, rather like Decca in Europe. However if a yacht is re-equipping, there really is no argument, for in my opinions GPS has everything going for it, not only in reliability and accuracy but in simplicity of operation as well.

Radar

I am completely convinced by radar and consider it a primary safety tool. There is an argument, which I do not subscribe to, which states that it is difficult to use, and in areas of good visibility not worth its weight or expense. We have found it a major help in navigation, particularly as a back up to Satnav when close to land. In recent years radars have undergone a major revolution, they are now cheaper, lighter, smaller and need less power to run them than their predecessors. The scanners also are now much smaller and lighter. The new sets use rasterscan with daylight viewing displays that eliminate those awkward hoods. Among the standard features on all sets are range rings, electronic bearing lines and interference rejection. Now radar can be interfaced via NMEA for visual display of compass bearings through GPS or Satnav, plus many other complex features.

I was particularly impressed with a new set that offered a feature called *synthetic afterglow*, which means in practice that a target will trail an afterglow tail. This can enable the operator to quickly assess other vessels movements. A problem with these sets is that they can be very power hungry, but this can be mitigated to some extent by switching to stand-by mode when not actually using the screen.

Familiarisation with the controls and the actual performance characteristics of a particular set may take a few hours of practise, but the actual tracking of targets, reading the working picture and plotting techniques are not difficult to learn.

Navtex

Navtex provides the yachtsmen with the latest weather and navigation information within the programmed sailing area. Currently it is available in most European waters and it is spreading world wide.

The principle is simple in that the receiver is tuned into the frequency 518 kHz and is left on to receive messages that are relayed from coastal stations in the programmed area selected; these are called Navarea. The Mediterranean has some coverage and more are planned for the future. All message are received in English and printed out as hard copy in order that they can be kept and referred to later if required.

Irrespective of the station selected and message requirements, all gale warnings, and search and rescue calls will be recorded. The system is an excellent one for yachtsmen, being reliable and automatic. There is no need to remember to 'switch on' for the weather forecasts – it's all printed out at regular intervals. We have found coverage is excellent right round the coast to Gibraltar. Then it gets a little sparse until the French coast is reached, but more stations are due on air so coverage right through the Mediterranean, should be available in the foreseeable future.

Chapter 14
Emergencies and Safety at Sea

Panic Bags, Fire, Life-rafts, Radio & EPIRBs, First-aid.

Panic Bags

Every offshore boat should have one in case of abandoning ship and every skipper will have his own ideas on what it should contain.

Firstly, I believe the panic bag should not be too big or too heavy; if it is, it will be difficult to stow in an accessible position. There are a number of excellent screw top containers on the market that are very good, but I favour a watertight duffle bag, for its soft flexible shape will allow it to fit into a variety of positions on board. We place ours on the engine housing just by the companion-way ladder, where it can be reached easily from below or from the cockpit.

For a cruise to the Mediterranean the contents need not be as comprehensive as for an Atlantic crossing. On the shorter trip, due to the frequency of shipping, help should be on hand much quicker in the event of an emergency.

As a basic guide, assuming the bag might have to support a crew of four for up to four days.Include:

Flares: 8 parachutes, 4 white hand-held, 4 red hand-held, 6 smokes. A Verey pistol with a dozen each of red and white cartridges.
Torches: 2 with a pack of spare batteries.
Water: 4 litres in plastic bottles.
Food: 4 large bars of chocolate or other high energy snacks, wrapped in cling film.
1 tin of glucose sweets.
24 Mars or Tracker bars.
2 tins of baked beans.
2 tins of tuna fish.
2 tins of peaches or other fruit.
2 tin openers.

An EPERB.
4 space blankets (one for each crew member).
Seasickness pills and ointment for salt-water sores.
Small first-aid kit.
Pocket knife.
Towels.
Plus any warm clothes there is space for.

This makes a pretty large parcel to handle, so it is doubly important that it can be reached quickly in the case of an emergency.

Other items to include separately, or if there is room in the bag, would be a hand held VHF, the crew's passports and ship's papers, money and credit cards plus extra food and water.

The life-raft is likely to be most uncomfortable. It will be wet and cramped, so items such as cards, books, paper and pencils etc., are worth taking to keep up morale.

Lights at Night

The 'International Regulations for Preventing Collisions at Sea' have been carefully compiled for the benefit of those who use the sea, both large and small. Lights convey a visual picture that ensures safe passage for ships and it is beholden to us yachtsmen to understand and comply with these regulations at all times. I have little patience with those who believe because we are small and are sailing on a sea with perhaps little traffic, we can get by without showing the correct lights.

The usual excuse is one of shortage of power, but this is no excuse at all, other than perhaps in an emergency. If a yacht is unable to maintain enough power in her batteries under normal conditions to keep her navigation lights burning, then in my book, she is not seaworthy and should not be at sea in the first place.

The actual regulations are not complex and the light patterns for various ships, tugs etc. can be quickly learnt.

Safety and Emergencies

When planning a long trip that includes open water passage making, the safety of the yacht and crew are uppermost in one's mind. Quite obviously, every possible contingency cannot be allowed for, but the broad principals of safety for the ship should be carefully considered, well before departure.

Safety

The ultimate survival of your yacht can easily be at risk by such small things as a loose jubilee clip on a through-hull fitting or the breakdown of a neglected bilge pump. Everything that has a direct bearing on the watertight integrity of the ship should be very carefully checked over, serviced and replaced if at all suspect.

Gas is a dangerous fuel that has caused many accidents, so check all connections carefully for leaks, and don't forget the piping around the cooker itself. This piping, due to the fact that the cooker swings in gimbals has to be flexible, so it is more liable to be worn and faulty than rigid copper piping. A sensible precaution is to install a detector in the bilges to warn of any build up of gas. Site the gas bottles in a locker on deck, with drainage holes in order to drain any gas leaks overboard.

Flares It's seamanlike to ensure all the flares are in date and that there are plenty of them. Large parachute flares are essential, as are the hand held smokes. I would suggest a minimum requirement for offshore cruising would be 10 parachutes

and 10 smoke, both white and red/orange. Keep them all in a watertight container close to the panic bag location. If in a distress situation, fire the parachute flares in pairs, a minute or two apart.

Fire-Extinguishers Locate the extinguishers and asbestos blanket a little away from the areas of maximum fire risk. So often one sees a fire-extinguisher clipped to a bulkhead directly behind the cooker, then if a fire starts in the galley, to get at the extinguisher would mean reaching over the flames, which makes no sense at all.

Position the extinguishers high up from the bilges, and have at least one for each main cabin. Locate a large one close to the main hatch so that it is easily reached without having to go below, then in an emergency you can start fighting the fire from the cockpit.

Types of Extinguisher

Dry powder extinguishers are the usual type found on boats, as they are efficient on most types of fires. The weakness with powder is that it does not have the capacity to cool, and if anything, traps the heat in, so once the fire is out, water should be sprayed onto the area to cool it down.

Carbon Dioxide is first class for engine-room fires as it displaces the air. Extinguishers can be purchased that are automatic in operation. If the temperature reaches a given level, the extinguisher is triggered.

Foam is another very good fire fighter, but it does make a terrible mess that is difficult to clear up.

Water of course is good to reduce temperatures, but should not be used on oil fires as it spreads the flames, or on electrical fires, if a mass of short-circuits is to be avoided.

Fuel

Diesel fuel is not a serious fire risk in itself

as its vaporising temperature is high and at normal temperatures it is difficult to ignite.

Petrol on the other hand is extremely dangerous and should be handled with great care. Always store petrol in metal cans that are in good condition (watch out for rusty seams). Ensure they are chocked off well, against the ships motion, so that they cannot move in a seaway. Really the best place for them to live on passage, is in a heavy plastic bag to keep out sea-water, then securely lashed to the pushpit on deck.

On Deck

Every yacht should have secure lifelines running round the side decks; these should be supplemented with jackstays to allow the crew to clip on their harnesses when moving around the deck. The jackstay terminal points should be well anchored and supported with pads under the deck as the shock loads of a heavy body being brought up short by a harness can be substantial.

At least two independent life buoys complete with lights should be positioned close to the cockpit on the rail or pushpit, and at least one of these should also have attached to it a dan-buoy with light and drogue. There is a very good case for equipping the yacht with a specific piece of equipment that will directly help the rescue of a crew member in a man overboard situation. The new Jon Buoy Mk III™ is a sophisticated piece of equipment using an inflatable miniature raft with slings for recovering a man from over the side.

Personal Safety

The first requirement must be for every crew member to have a personal, modern harness, preferably fitted with non-carbine type hooks. Also, a life-jacket should be available with personal safety light attached. It is a good idea to apply some retro-reflective tape to life-jackets, harnesses and oilskins. This tape can be stitched on, or in some cases it is self- adhesive. The result is that it increases the chance of being seen (and rescued) at night to a marked degree. There is some active development going on with miniature rescue radio transmitters that can be carried on the person. These are automatically activated when immersed in water and the transmissions are picked up by an onboard receiver, so giving a direct bearing to the crew member to be rescued.

When entering the cockpit it makes good sense to get the crew to clip the harness onto the pad eyes whilst still in the safety of the companion-way. It takes just a moment for the boat to lurch and the unsecured crew member can be flipped out of the cockpit in a moment. The same principle should also be applied to jackstays. Before attempting to climb onto the deck to go forward, get into the routine of clipping on before leaving the cockpit. It's a particularly vulnerable time as you are standing up on the cockpit seat or coaming and your centre of gravity is that much higher. I make no fuss about moving around on deck for I usually crawl around on hands and knees or on my bottom, that way my centre of gravity is low and I have two hands to hold on with and two feet to brace myself .

Emergencies

The major emergency is for the boat to sink, either as a result of a collision, explosion or other rupture of the hull. The sea is littered with all manner of debris and it's not unknown for yachts to sink after colliding with some object floating just below or on the surface. The main priority will be to get the crew safely off into a rescue craft of some sort, then to alert someone, anyone, to come and pick you up.

Life-rafts

There is a lot of bunkum talked about the durability of life-rafts at sea. I hear comments from some people that they are useless in heavy seas, they would never have one, etc. but you can bet that they are not the people who venture offshore, and if they do without some sort of life-raft, they are extremely foolish people. The prospect of a boat sinking, without the back-up of a life-raft or some other specialist craft to launch, fills me with horror.

It is true that the traditional life-raft has its shortcomings – it can fail to inflate, but regular servicing should circumvent that. It can capsize in heavy seas – but at least you can get it upright again using the straps built into the lower tubes for this purpose. It certainly won't be comfortable, but on the other hand you won't be dead either. A great deal of development has gone into life-raft design over the last few years and they are generally far more stable than they were. They are your last ditch survival aid so make sure they are in perfect condition before you start an offshore voyage.

Just before departure have the raft serviced, and ensure the agent carrying out the work is qualified. The points to check are: It stays inflated for at least twenty four hours at high pressure. The inflation cylinder is in good condition, clean dry and full. The hand inflator used for topping up is in perfect working order. Seams and all structural elements are checked. Survival kit complete and in good condition (replace the torch batteries as a matter of course). Flares are in date. Canister or valise is in sound condition. Service card updated and stamped.

When storing the raft on board it should be protected from heavy seas and rain. Something that is not generally understood is that raft containers are not always fully waterproof. Mount canisters with the drainage holes facing downwards. Keep a valise in a locker or similarly protected position, but ensure it can be got at quickly in the case of an emergency. Don't store them on end for any length of time, and protect them from heavy handling and knocks.

Some yachtsmen swear by the adaption of the Tinker Tramp inflatable dinghy into a survival craft. This is carried out by adding an inflatable canopy over the complete dinghy. The advantage of this craft is that presumably in moderate weather, the canopy can be removed and the little craft sailed to the safety of the nearest land.

Other modifications to existing 'solid' dinghies centre around a large inflatable collar mounted around the boat's gunwales so rendering it unsinkable. In principle, if the conversion work is carried out to a very high standard it should stand up to the rigours of stormy seas, but I am less sure about the inmates being tossed around in a hard shell. Another consideration is that it would take up a great deal of space on deck, and be difficult to launch in an emergency if a high sea was running.

The other important consideration in an emergency at sea is the alerting of other ships and rescue services to your plight.

VHF and SSB Radio

A marine radio is without doubt a wonderful asset to have on board. The SSB or Single Side Band in its best forms is a world wide communication radio that works through shore based telecommunications stations. In the UK this is via Portishead and the Caribbean, Barbados. It is expensive to buy and install, but once set up and running should prove reliable over a long period.

VHF is a short distance radio that is effective to around twenty five to fifty miles depending on local conditions. It is particularly useful for ship to ship communication in an emergency, provided of course the ship one is calling is within range.

EPIRBs short for Emergency Position Indicating Radio Beacons are a necessary piece of equipment to have on board for any yacht venturing far offshore. They have been developed over the years, so that now they are world wide in their coverage. There are two types available operating on different frequencies. The inexpensive, small sets working on 121.5 Mhz are constantly monitored by aircraft flying normal inter-coastal routes, in consequence the waters of the Atlantic are well covered.

The other more sophisticated set-up works on the 406 Mhz frequency and takes advantage of the new professional COSPAS-SARSAT systems which claim advantages such as global coverage and improved location accuracy down to two and a half miles. The way this system works is that once triggered, the EPIRB sends out a signal that is intercepted by one of the five polar orbiting satellites, (three Russian, called COSPAS and two American called SARSAT). The signal is then returned to earth where it is received by a Local User Terminal (LUT). It is again is processed and finally passed to the control centre which alerts a local rescue service.

This sophistication and reliability is of course more expensive that the simpler 121.5Mhz sets but for anyone contemplating cruising in areas not covered by overflying commercial aircraft, it would be money well spent. In addition, there are now sets available that can be deck mounted so that they automatically trigger themselves, some even float-free when a quick release mechanism is activated.

There have been very many successful rescues documented over the years where the rescue services have been alerted via an EPIRB. They are not always reliable however, so a close watch should be kept on battery state, and care should be taken, particularly with the less expensive models, to ensure they are positioned away from an area that can get wet. There is a good case for keeping them in the Panic Bag, provided it can float and is watertight.

Illness at Sea

To anyone like me, that has the greatest difficulty in coming to terms with the sight of blood in any form, trying to take on the role of a medic in an emergency is a problem. However it is very necessary for the skipper or some other crew member to understand even the basic rudiments of first-aid, so that should one of the crew fall ill or have an accident there is someone on board who has some knowledge of the subject.

Certainly before an ocean passage is contemplated it is wise for one member of the crew to take some instruction on the basics of first-aid.

The usual injuries on board a boat fall into the cut, bruise or broken bone categories, which thankfully are not life threatening and can usually be treated with the aid of a good medical book.

On longish trips of say two weeks or more it can be worrying if a crew member falls ill or if an injury develops complications, then the only course of action is to try to summon help from a passing ship, or in real emergency via an aircraft, by activating an EPIRB.

A medical check-up before departure is a sensible precaution, also a visit to the dentist, for safety's sake. Can you imagine three weeks of severe toothache? Other than that, anyone in reasonably good health should not have too many worries. Sailing is a very healthy pastime and the chances are that you will return far fitter than when you left.

First-aid

I recommend the first-aid kit is as comprehensive as you can make it, and if some of the contents are used during the cruise, remember to replace them on arrival just as soon an possible.

Your local doctor will give you a list of the main medical necessities, and as the yacht will be out of UK waters, such things as antibiotics can be prescribed. We also ship a quantity of hypodermic syringes on board just in case they are needed in some sort of out of the way place, where cross infection might be a problem. To accompany these, we have a letter from the medical authority explaining the reasons for these needles, as we certainly don't wish to end up trying to explain in a foreign language to some drug enforcement officer that we are not all junkies or drug runners!

The following is a list of medical supplies we carry on board, it is by no means exhaustive but forms the basis of a good medical kit.

Wounds, cuts & abrasions:
Antiseptic cream and liquid (Savlon™)
Antibiotic powder (Cicatrin™) aerosol
Sterile gauze swabs (10 pkts)
Distilled Witch Hazel
Waterproof plaster (2 inch and 1 inch adhesive)
General plasters (2 pkts)
White open-wound bandage in 1, 2 and 3 inch widths
Crepe bandages 2 inch
Sterile skin closures
Tincture of Iodine

Breakages and Sprains:
Splints
Elasticated tubular support bandage (2 packs)
Burns:
Burneze™

Iodine dry powder spray (Disadine™ DP)
Burn dressings
Bites and Stings:
Autan™ cream or stick (repellant)
Antihistamine cream
Eyes:
Sterilized eye pads
Eye drops (Fucithalmic™)
Ears:
Sotradex™ ear drops
Athletes Foot:
Mycil™ ointment
Seasickness:
Stugeron™ tablets (Cinnarizine 15mg)
Chapped Hands and Lips:
Lypsyl™ – Uvistat™ 5g
Sunburn:
After Sun™ soothing milk or cream
Chest or General Infections:
Antibiotic such as Amoxil™ 250mg or any wide spectrum antibiotic
Pain-killers:
Strong: Co-Dydromol™ or Fortal™
Medium: Codeine
Light: Aspirin or Veganin
General:
Stainless steel scissors, a box of safety pins, hypo syringes, stainless steel forceps, thermometer, antrid tables, finger stalls, eye bath, cotton wool & buds.

Plus a good book on first-aid and simple doctoring at sea.

Chapter 15
Food and Keeping the Crew Happy

Fresh Produce, Dehydrated Food, Cereals, Galley Equipment Offshore Recipes.

When cruising we are great believers in conceding as little as possible of the normal standard enjoyed in shore life.

Food can be one of the greatest pleasures in sailing offshore and I can see little point in enduring the hardships of monotonous diets, made up from tinned stews and bully beef sandwiches that seem to feature in so many boat menus.

Apart from the enjoyment of eating good healthy food, the practical considerations of fuelling the body with a balanced diet should not be ignored. On a lengthy cruise where the boat will be away from fresh food supplies for perhaps twenty days or more, there may be particular problems unless the ship is endowed with a large freezer (let's hope it won't break down). To overcome this, careful thought has to be given to supplying a diet that will be nourishing as well as interesting to those who have to eat it.

Fresh Produce
Fresh meat, fish, vegetables and fruit are most important and should be stocked up just before leaving. A normal yacht's refrigerator will keep meat and fish fresh and in edible condition for three to four days. Pre-cooked meals should be consumed within two days to be on the safe side. With careful planning it is possible to supply the crew with fresh or pre-cooked meals for the first four to five days without too much difficulty. The rest of the time the meals will have to be from dehydrated or tinned sources. Again, if imagination is used in selecting ingredients the actual meals can be delicious and varied. On principle I refuse to have a single can of tinned stewing steak

on board, it's now quite offensive to my palate after suffering many years of racing, where the basis of so many main meals seemed to stem from this product.

Fresh Fruit and Vegetables
A good plan is to ship enough of these on board to last the whole trip. With fruit, carefully select unblemished examples and try to vary the ripeness from immature to nearly ripe when buying. This is not as difficult as it sounds, as most supermarket managers are quite helpful, and if requests are made for unripe fruit, they will obtain some from their stores. Soft fruits like pears, bananas, grapes, peaches etc. should be stored in nets to ensure good air circulation. Oranges and apples keep well and provided any bad fruits are regularly removed should last for three weeks without too many going bad. With vegetables, leaf produce such as lettuce, watercress etc. have only a short life, but root vegetables, carrots, potatoes and of course onions last very well. To keep some greenery going we sow some mustard and cress on damp towels or in old punnets. These little 'gardens' are very simple to tend and do provide a welcome addition to the menus as the cruise continues. Other 'sowings' can be made with alfalfa. This will give you a reasonable crop in two to three days. Mung seed is also worth trying and will be edible in about five days from sowing.

Eggs
Eggs are a most useful and versatile food offshore. Contrary to some opinion, they will in fact last very well without treatment of the shells, provided they are freshly laid and

have not been chilled. Keep them in their fibre boxes and place them somewhere cool, such as on the water tanks or in the bilges. They will keep quite fresh and edible for up to six weeks or more if stored in this way.

Bread

Wholemeal and granary bread can be kept fresh for up to ten days. The sliced variety a little less. White becomes stale much quicker. We have used with great success the part-baked bread obtainable from most supermarkets. It is simple to prepare, just pop it into a heated oven for a few minutes and wonderful freshly baked bread will emerge that is quite delicious. It is so good we use it as the mainstay of our bread supplies now when cruising, and although not cheap, it makes a super addition to the meals on board. For some reason the marked 'use-by' dates are quite short, but we have kept it in perfect condition for over two months on board without deterioration.

Butter and Cheese

Wrap butter in foil and keep it cool on the tanks or in the bilges, where it will stay edible for six weeks or so. Cheese can be kept fresh for months if vacuum packed. Before departure I grate a quantity of Cheddar and seal it in airtight bags. It will last for up to two weeks like this and is a great convenience for sandwiches and use in cooking.

Pasta, Rice, Pulses

These become the mainstay of many meals when cruising and thankfully they keep wonderfully well with the minimum of attention. We store ours in a selection of airtight plastic containers that are damp and bug proof.

Herbs

Ship aboard a selection of dried herbs, they can liven up the most ordinary meals. On Mishka we have a special herb rack to keep them safe and it is positioned in a place of honour just by the galley stove. Refill packs can be purchased almost anywhere and we keep a selection of the most used ones such as parsley, mint and mixed-herbs as a stand-by. Garlic is a wonderful herb and will keep well if stored in a dry airy place. Garlic paste is available in tubes and perhaps is rather more convenient to use on board.

Meat

It is possible to buy vacuum packed meats from some butchers and supermarkets, and provided the seals are not broken they will last safely for a week or two. I am a little nervous about meat in general and take particular care when using it. For some reason lamb will keep fresh longer than other meats; but if it starts to get a little 'high' cook it immediately. A smoked ham is well worth keeping on board as it will keep almost indefinitely, the same applies to whole salami – wonderful for spicing up omelettes and salads.

Dehydrated Meals

There are a good selection of dehydrated meals available through stores and specialists shops and are excellent as standbys. Some yachtsmen use them all the time, but in my opinion they tend to be mushy and rather lack flavour, and are also expensive. Recently I came across a home hydrator that seemed to be the answer to long term storage of food on board. The principal is that the food to be dried is placed onto a series of stacking trays, which are then placed on top of a metal base, housing a fan. The unit is covered with a lid, the thermo-stat is set and once the electricity is switched on the hydrator is left to its own devices for about eighteen to twenty four hours. The standard unit can dehydrate about six to eight pounds of food at a time and it is claimed that almost anything can be dried from fresh fruit and meats to complete meals.

Tinned Food

Not the ideal food to live on, but inevitable for most of us as basic fare once the fresh supplies have run out. We try to keep the consumption of tinned food down as low as possible and to stock the more unusual varieties. The good basics that are very useful in a multitude of dishes, are such things as:

Tomatoes, beans, salmon, tuna, sardines etc.

Peas (freeze dried are best), sweet corn and other vegetables.

Fruit (in natural juices not syrup).

Also tins of sausages, paté, chicken, ham and turkey come in very handy for risottos etc.

Spices and Pickles, Soups etc.

Dried packaged soup is vastly superior to the tinned variety and if selected carefully the quick heat types only take a few minutes to prepare.

Spices in various flavours such as chilli, cinnamon, cloves, cayenne pepper come in very handy and it's worthwhile stocking a number of different pickles, not forgetting tomato and Worcester sauces. Dried mustard keeps well, lemon juice in bottles or 'Jif™' containers. Parmesan cheese in a block for pasta dishes and a variety of stock cubes are excellent for more complex menus. Curry powder and paste is a necessity, as is long-life and dried powder milk. Some jars of peanut butter and Marmite™ for use in sandwiches are well worth a place to spice up meals.

Cereals

There are a wide selection of breakfast cereals available, most of which are of the sweet non real-food variety. We make up our own breakfast mixture and use it also as an anytime meal, stand-by. Sometimes one needs a very simple nourishing meal quickly and this fits the bill very well. We mix it from the following ingredients but the actual mixture can be varied to taste.

6 parts unsweetened muesli
4 parts bran flakes
4 parts Special K™ flakes
2 parts bran buds
2 parts All-Bran™

To this we add chopped dried figs, apricots and prunes together with dried raisins. The whole lot is mixed up together and bagged in airtight containers. It is our basic breakfast cereal and quick meal stand-by for watches. Not only is it delicious, it also provides a sensible basis of dietary fibre which is doubly important on a boat where the meals tend to be rather mushy in nature.

Snacks

Although most of these have limited food value, their importance is greatly magnified by the lack of other shore based 'treats' available to us sailors. We tend to stock a small quantity but a wide variety to keep interest up, and try to avoid too many chocolate based things. I suppose on most peoples lists Mars™ bars would be near the top, but as the boat is heading for the hot weather, these are augmented by products such as Tracker™ and muesli bars. Savoury snacks are also much valued as a break from the sweet variety, and we always have a good selection of dried fruit available.

Galley Equipment

A word or two on galley equipment might be useful at this point. Depending on the culinary skill of the cook, the equipment has of necessity to be of a simple kind, due to the lack of available space in a typical yachts galley.

Pressure cooker, wonderful for all-in-one type stews and general cooking of the meat and vegetable type of meals.

At least two *frying pans* of the deep sided variety – non stick if possible (fried food should be avoided, as high fat meals lead to dicky tummies) The reason for recommending two is that I have an excellent recipe that uses two pans.

Three saucepans of varying sizes.

One *kettle*.

A *steaming* net that fits into a saucepan.

Two or more *ovenware dishes* (stainless steel are best).

A *baking tray*.

One or two *potato peelers*.

Some good quality kitchen *knifes* plus *sharpener* and *egg whisk*.

A *casserole*.

Three *tin-openers*.

A *bottle-opener* and *cork-screw*.

A very good *chopping-board* that can be fixed or locked in position, galley *scissors, spatulas, spoons, colander*, stainless steel *grater*, as much *kitchen paper* as you can store of the high absorbency type, plus *aluminium foil* and *cling film*.

Offshore recipes that are good to eat

I don't propose to produce a recipe book here, but there are one or two well tried recipes that have proved very popular with crews, they are simple to make and are delicious and healthy to eat.

Mishka's Omelette

(This is the one that uses two frying pans, one slightly smaller than the other.)

Will feed 4 hungry crew:

Ingredients
8 eggs
Large tin of sweet corn
About 6 medium potatoes or equivalent of Smash™
A dozen slices of salami and a tin of sliced ham
A handful of freeze dried peas such as 'Surprise™' Peas
One large onion
Mixed-herbs
A little salt & pepper, cooking oil and 1 large garlic bulb

Method
Peel and boil the potatoes until soft; then slice them up into 'rounds'. Chop up the onion and fry in oil, using the smaller of the two frying pans. Add the garlic and potatoes then cook until starting to brown.

Chop up the salami or ham and add to the mixture together with the corn. Heat through well and tamp down a little to make the whole mixture solid (it should be about two inches thick).

Whisk the eggs thoroughly with the herbs and pour over the mixture. Cook until the bottom layer is well set and crispy brown. Now comes the clever part! The top of the mixture is still wet and uncooked, so bring forward the larger of the two frying pans, also slightly oiled. Place over the pan with the mixture in and tip the whole contents back into the big pan. Then cook for a couple of minutes until brown and set.

The result, is a really tasty and nourishing mega-omelette that will satisfy even the largest appetites.

Bistro Ham
(a really delicious meal for 4)
Slightly more complex, but well within the capabilities of the average hungry crew to prepare:

Ingredients
1 tin ham
4oz fresh, tinned or dried mushrooms
1 large onion
1 tin tomatoes
¼ pint milk
1 tablespoon Dijon mustard
2oz butter
¼ pint white wine
1 tblsp flour
1oz Parmesan cheese
¼ pt yogurt or cream (tinned will do)
Parsley, salt and pepper

Method
Butter an ovenproof dish and cut up slices of ham and arrange on the bottom. Spread mustard over the ham. Add sliced and chopped onion and tomatoes.
Melt a little butter in a saucepan, add mushrooms and wine and cook for three minutes, pour mixture over ham.
Again melt a little butter in pan and slowly blend the milk (off heat) and flour until thick, season. Add parsley and cream or yogurt (off heat) and pour the sauce over the ham, sprinkle with cheese and cook in moderate oven.
To bulk the meal up - cook and cream some potatoes with perhaps a tin of peas. Quite superb and really very easy to prepare.

Tuna & Mushroom Spaghetti
(serves 4 crew with big appetites)

Ingredients
1 large tin tuna fish
1 large onion
3oz butter
3oz mushrooms
Small carton or tin of cream or yogurt
2 tblsps tomato paste
Garlic, parsley, salt & pepper
Enough spaghetti for 4 crew

Method
Chop up mushrooms. Melt butter in pan, add chopped onion and drained tuna, stir over moderate heat until onion is just tender.
Add mushrooms, cream, tomato paste, herbs and seasoning. Bring to boil and serve over pre-cooked spaghetti or other pasta - wonderful and very filling.

Mealtimes & Keeping the Crew Fed
A routine for meals should be worked out well before departure to ensure everyone knows what is expected of them. A method we have worked out that suits our particular style of cruising is a simple one.

Breakfast – Everyone gets their own breakfast at the time they want it, so sleeping off-watch members will not be disturbed. Normally the basic menu centres around our special breakfast cereal, plus toast and marmalade with coffee. We rarely have a cooked breakfast offshore and never have one using the frying pan, as greasy food (plus having to wash up the frying pan) don't seem popular with us at sea.

Mid Morning – Coffee or beverage of some kind depending on the weather, plus biscuits (we carry lots of sweet and savoury biscuits in a wide variety to keep everyone interested).

Lunch – This is timed to coincide with the change in watches so that the whole crew are together to discuss the running of the ship. Usually the meal is prepared by the cook and is based on hot soup (we carry four different varieties to ring the changes) and sandwiches. The sandwiches are varied and a major meal in themselves. No thin slices of meat between a couple of bread slices for us! We use all sorts of mixes mainly thought up by my family to make this an interesting meal. Some of the favourites are:

Chopped ham
(tinned) with sliced apple or tinned pineapple with cucumber, lettuce and mayonnaise.

Peanut butter with Marmite
sliced tomato, cucumber and salad cream.

Grated cheese with pickle
lettuce, grated carrot & cucumber and mayonnaise.

Sliced chicken
(tinned) with mustard, any green salad, tomato and salad cream.

Mashed banana
with bottom slice of bread spread with strawberry jam, topped with grated bourneville chocolate.

Tuna fish
with lettuce or green salad, sliced tomato and sprinkled with grated cheese and salad cream.

Tea – a great tradition on our boats is afternoon tea and we invariably serve it about 1600 hours wherever we are. One of the standbys for cruising afternoon tea, is fruit cake and we put aboard one or two large catering size cakes for our mutual enjoyment.

Supper – This is the main meal of the day and we proceed it with a sundowner to choice accompanied by some savoury snacks. Again this is a time of day that all the crew are gathered together for a communal chat to discuss the days events, so whilst the cook is preparing the meal we enjoy a glass or two together.

It is an interesting fact that perhaps up to ninety percent of our meals are eaten in the cockpit (subject of course to weather conditions). I suppose this gathering to share a simple meal together is at the very heart of enjoyable cruising. Of course as the weather improves and the yacht tracks further south, so the gathering in the cockpit last even longer and become even more pleasurable.

Alcohol

We don't have any hard and fast rule about drinking on board, largely because all the crew we sail with are moderate drinkers and sensible people; they simply help themselves to whatever they want. I can see however that such a system could be abused and if we shipped a tippler on board as crew, we would insist that no alcohol of any quantity was consumed before taking over watch keeping duties.

A wonderful anchorage in the Balearics

Chapter 16
What to Expect When You Get There

Berthing, Mooring & Anchorages, Winds & Weather, Provisioning, Transport, Security, Bites, Burns & Bugs, Pollution, Tropical Waters, Swimming & Snorkelling, Care of the Underwater Hull.

Well there will certainly be a great deal of sunshine, particularly during the summer months. For Mediterranean cruising life tends to become less expensive (Italy and the South of France apart) the further east you cruise. It is also less populated than further west, and for yachts and their equipment requirements, less sophisticated. Travel will also be less convenient and from northern Europe rather more expensive.

Berthing, Mooring and Anchorages
New berthing techniques will have to be learnt, for in almost all areas, yachts are berthed bow or stern to, using the ship's anchors, or docking lines attached to mooring chains that run along the bottom.

The correct technique for mooring with a bow anchor, is to slowly motor up to the berth, drop an anchor, then going slowly astern, pay the cable out until the pontoon or berth is reached. Mooring lines are then taken aboard over the stern and made fast. In some marinas there is a mooring line attached to a ground chain, so that dropping an anchor is unnecessary. This simplifies the job a little, but it does mean the yacht has to motor into her berth completely, before any lines can be taken aboard. However, these techniques are soon mastered with a little

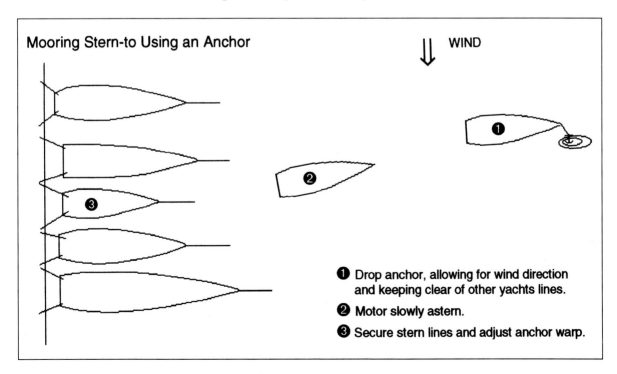

Mooring Stern-to Using an Anchor   ⇓ WIND

❶ Drop anchor, allowing for wind direction and keeping clear of other yachts lines.

❷ Motor slowly astern.

❸ Secure stern lines and adjust anchor warp.

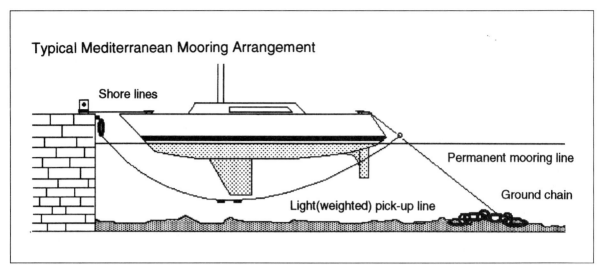

Typical Mediterranean Mooring Arrangement

Shore lines

Permanent mooring line

Ground chain

Light(weighted) pick-up line

Getting Ashore

Efficient and safe but difficult to stow.

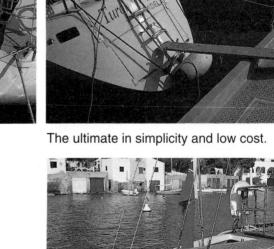

The ultimate in simplicity and low cost.

A good system that folds for easy stowage.

Mishka's elegant answer.

Some boats prefer to face bows-on to the quay and have adapted their pulpits accordingly

practice, fortunately there is usually little or no cross wind to complicate issues, and of course no tides to worry about. The important thing to remember, is to approach the berth slowly, with plenty of fenders down, so, if the initial run is incorrect, no damage will be done, and you can try again. We have found almost invariably that cheerful help seems to be on hand if things get awkward, a not unheard of occurrence with us, due to *Mishka's* unpredictability when motoring astern.

Anchoring in the multitude of bays and Calas that are so much a feature of Mediterranean sailing is usually straightforward although, bottoms can be steep-to and sometimes unreliable, due to weed covered rock. Depths are considerable almost everywhere, so that boats can sail very close to land without danger, which makes cruising so much more interesting.

Winds and Weather

The weather in the Mediterranean can come as a surprise; for, so many people imagine it to be largely calm with light winds blowing out of a clear blue sky. It is true that this can be the case on many days, but it can, and does blow hard at times, and on

occasion, very hard indeed. Happily for the most part the weather is predictable, usually with light wind in the mornings, breezy afternoons once the land heats up and starts the sea breezes going. Then as evening shadows fall the wind drops and the nights remain soft and calm.

In general the winds tend to be thermally generated, therefore in settled conditions they can be relied on to conform to a pattern, but the Mediterranean can also be dominated by other influences. The Sahara has its effect, in that during summer, the heat it generates controls the surrounding area to a large extent. If winds develop from the South they can carry deposits of sand with them for hundred of miles into the Mediterranean area.

Thanks to the Azores High, the Atlantic depressions that effect the weather so much in northern Europe are effectively blocked off from the Mediterranean, so giving the area a stable cyclonic environment with settled pressure for most of the summer.

Southern France can on occasion be rather different, for it does tend to be windless for many days in summer, only to be torn apart by the roaring winds of the Mistral as it rushes down the Rhone Valley

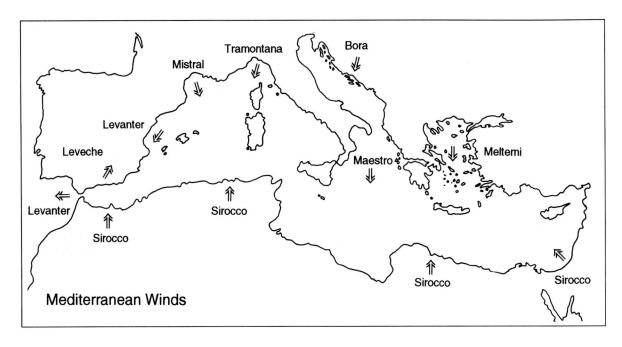

Mediterranean Winds

The main winds of the Mediterranean - local names.

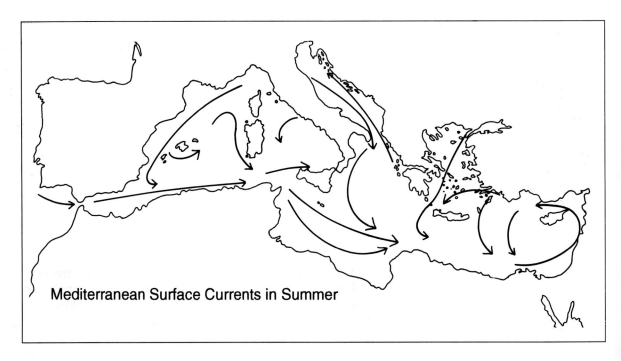

Mediterranean Surface Currents in Summer

In general the surface currents flow anti-clockwise and are normally weak in strength, however they can increase around headlands and between islands. Strong winds blowing against the current can cause steep waves and broken water at times.

out to sea. The effects of the Mistral can be far reaching, extending far beyond the coast of France as it spills out into the Mediterranean to the coast of Spain and sometimes even effecting the weather of the Balearic Islands. The Meltemi is another wind that can reach gale force at times. It is largely felt in Turkey and the Greek Islands, blowing from the North-East during late July to September. This can make cruising difficult if one is headed north, therefore spring is the best time for sailing into the Aegean Sea; then the idea would be to cruise southwards as the season lengthens.

The Sirocco is a southerly wind that develops in the dry southern deserts from such countries as Egypt, Algeria and Morocco; it can occur in fact all down the southern seaboard of the Mediterranean. Usually the wind is 'dirty', with sand particles which can smother a yacht and stain sails and covers. If this happens the sand can be washed off effectively with water, but if left for any length of time the sun will bake it on, then it will be the very devil to remove.

We have been caught in the Levanter that tears through the Strait of Gibraltar and wreaks havoc along the southern Spanish coast. It can blow very strongly at times and effectively 'shuts the gate' at Gib so that only ships and very large motor yachts have any chance of getting through. These major winds are in the main relatively uncommon in the summer and usually well forecast.

Contrary to common belief, we have found very good sailing in the Mediterranean, with fresh winds occurring perhaps two days out of every six or seven. What is a feature however, is the dramatic effect high land has on wind strengths and their direction, so care should be exercised when approaching islands and high landfalls. In such circumstance we have experienced a Force 5 suddenly gust to Force 7 or even 8 with little or no warning other than a slightly ruffled sea to windward.

Summer sailing is, on the whole, delightfully warm and undramatic, provided a weather eye is cocked for thunderstorms and the squalls that usually accompany them. It is sensible to have a powerful, reliable engine to maintain cruising schedules as the evenings and nights can be calm, but other than this, the sailing can be wonderful with deep blue seas, fresh warm winds and sunny skies.

The best of the sailing season really is from early April until late October in most areas, although it can be extended a little if care is exercised and weather forecasts are heeded. In winter the whole region can become quite cold with bone chilling winds and even snow in some places. If the boat is left in commission during the winter period, a good safe harbour is necessary, with heavy warps and plenty of anti-chafe material around the fairleads. My preference is for a roaring log fire and a cosy sitting room in England, if I can't be in the Caribbean that is.

Provisioning

In general, local foods are the best and cheapest wherever you are. Imported goods are invariably expensive and on occasion difficult to obtain. As a rule we always try to avoid the 'marina' shop and try to find the markets or small stores where the locals buy their food. Some countries, notably in the western Mediterranean have very good supermarkets and it's a simple task to stock up for a week or so. In other places you must seek out the best shops which is not always as simple as it sounds, as they can be miles from a chosen secluded anchorage. Shopping hours vary, but as a general rule most open for business about 0800hrs to mid-day then close for siesta until 1500hrs, when they open again until late evening.

Fresh fruit and vegetables vary a great deal in quality, price and availability. The places to obtain the best quality are in the local markets.

Cheese is produced on a local basis virtually throughout the whole region, and very good it is too, normally being reasonably priced and of excellent quality.

Wine of course is produced almost everywhere and can be first of class quality and reasonable in price – sometimes considerably cheaper than water!

Fish should be plentiful and inexpensive, but sadly this is not always the case. Portugal, Spain, Italy and Turkey for example are good for fish whereas we have found with Yugoslavia fish is comparatively scarce and can be expensive. It is best to visit local fish markets or to buy direct from the local fishermen on the quay.

Meat is plentiful and inexpensive if the local product is purchased. If your palate demands other meat then be prepared to pay for it. Local sausages, despite the look of some of them, can be excellent and particularly good for barbecues; if in doubt choose the large frankfurters which again can be first class.

In general then, provisioning is usually straightforward with most things being freely available. You may have to get used to local products that don't have a familiar taste such as milk and cheeses, but this is a small price to pay for such varied foods and wines.

Water can be a problem at times, but we overcome this, largely by stocking up with fresh bottled water wherever convenient to do so.

Transport

Local transport can vary from excellent to chaotic, but it's usually fun if you have plenty of time at your disposal. We have found that taxis are available almost everywhere and usually at a sensible fare (always ask the price first). It sometimes pays to use a taxi to get provisioned-up at a local town's supermarket once a week, rather than pay over the odds at the small local store, particularly if its part of a marina complex.

One of our greatest assets is a couple of folding bicycles which we keep on board. The ones we selected are made by Strida, but there are many others that serve equally well. I feel it is worth accepting the extra cost to have the machines made from non-corrosive materials, for normal steel will soon develop rust and look shabby in no time. The extra freedom these little machines have given us has to be experienced to be believed. They are in constant use for shopping, sightseeing or just used for exercise on those lazy days in harbour. They have proven very popular with our family and guests, some even volunteering to cycle to the shops just for the fun of it!

Marina's and Security

The quality of marinas is very high in most major Mediterranean yachting centres, usually they have excellent facilities, are clean and generally very welcoming. In spite of the popularity of our sport, we have never been turned away from any marina even in the high season (July & August), but don't expect them to be cheap. In general, France, Spain and the Balearics are in line with middle priced south coast of England marinas. Italy and the more expensive parts of southern France, together with Sardinia and Corsica can be more expensive depending on how fashionable the chosen location is. Having said this, there are some smaller marina's that are more reasonably priced, but usually are less conveniently positioned along the coast line.

In addition, there are countless small anchorages available, sometimes adjacent to a tourist centre with local provisioning facilities. These anchorages are the very essence of Mediterranean cruising and we use them all the time, other than the odd marina visit to fill up with water or fuel and possibly stock up with food. Unlike the harbours and anchorages in northern Europe, in our experience a charge is never

A fairly typical Mediterranean marina with good lifting facilities

If the boat is being left unattended for a long period some form of secure mooring should be arranged.
The photo shows a short length of chain shackled to a large spliced eye. The nylon warp is protected from chafe where it enters the fairlead on the quarter.

made, the authorities seem to welcome yachtsmen, and show them every courtesy.

Security

To date we have not had any problem with theft, due perhaps to the fact that we try not to leave our boat in an exposed position for any length of time. It's common sense not to tempt providence, so we lock up securely when leaving the boat to go ashore. We have also selected a very safe marina when leaving our boat for long periods between our bouts of cruising, although rather more expensive than other places, the security system seems good, so we feel in the long run it's good insurance.

Theft and vandalism is unfortunately on the increase wherever you go, and in some areas of the Mediterranean and Caribbean it is rife. The best advice is to thoroughly check out the local situation with other yachtsmen, who live or have visited, before leaving your boat there. Places vary a great deal in security mindedness and a call to your insurance broker may also help in the selection of a wintering hole. Dinghies, outboards and loose deck equipment seem to be the things that 'go missing' most frequently and are usually the result of the owner being careless about locking them up. Most thefts seem to be of the petty variety born of opportunism rather than organised crime, so to a large degree the cure lies with us, to remove the temptation in the first place.

Bites, Burns and Unwanted Visitors

The familiar mosquito is unfortunately prevalent in many Mediterranean anchorages at certain times of the year, and can be really annoying at night. The time honoured methods of keeping them at bay are still probably the best but boats pose particular problems that make skeet control more difficult.

Netting is of course excellent but troublesome to erect every night. Many people advocate individual screens for all the hatches and opening skylights, but in my experience this cuts down too much on ventilation, when cooling breezes through the boat can make all the difference between sleeping or lying awake in a hot stuffy atmosphere. Mosquito coils which are lit and then smoulder away to make an insect repellent smoke are effective if you enjoy living in a foggy atmosphere, and anyway if there is a breeze blowing through the yacht it soon dispels, leaving the occupant with a headache and still liable to being bitten.

I feel the best method, other than rigging up full netting over every bunk, is to use the anti skeet creams based on oil of citronella. There are a number of good ones on the market such as Autan™, Jungle Juice™ etc. They have a very strong smell which some people (unlike the mosquito) don't object to, but in general are pretty effective. They don't do much for one's sex life, but at least enable you to get to sleep without being bitten.

We have tried another little gadget that seems to work well. It's a battery powered light that emits a blue glow from a cathode ray type of tube, which seems to attract insects, we call it our 'light zapper', for when insects fly into the light to be electrocuted, it emits a most satisfied zapping sound.

Skeet bites effect people in varying degrees. In all probability we all get bitten, but some of us react more violently than others. There are a number of preparations available that are supposed to help with the irritation, but we have found that for really nasty bites Wasp Eze™ is good, and calamine lotion helps in keeping the bitten areas cool. Dusk is the real danger time, particularly when at anchor, which is particularly annoying as this is usually the time when the crew are relaxing in the cockpit enjoying a meal after a satisfying days sail. For some reason, mosquito's always seem to fly against the wind, so if the yacht is anchored with an onshore breeze, the infection is likely to be greater – so get the anti skeet creams on early.

Most human beings have a deep rooted fear of snakes. It is true that there are some in parts of the Mediterranean, but very few incidents of bites have been recorded. The snake is basically a timid creature and will not attack a human unless it is frightened or injured in some way. The best way of avoiding them, is to wear shoes and socks when ashore in country that is a likely snake habitat. Also make plenty of noise when walking around. The vibration caused by heavy footfalls should give them plenty of warning and they will keep clear. If you are very concerned, carry a stick to flick them gently out of your path.

Burns can be a problem, and if there is a lack of prudence with the crew concerning sunbathing, it is likely to be something that will concern the 'doctor' on board on a regular basis. The sun can be very hot, and in general it is wise to keep out of it as much as possible, particularly in the hours around noon. A wide brimmed hat, with chin strap and light cotton shirts are the best apparel for sailing. If there is a crew member on board that has fair skin try to persuade them to wear a pair of light cotton pyjamas during the hottest part of the day, these are wonderfully practical and most comfortable to wear.

Don't swim or snorkel for long periods without some protection for the back. We always wear a T-shirt if staying in the water for sometime. Sun glasses are also very useful, for the reflected glare from the bright light on the water can effect some people quite badly. Polaroid glasses are excellent in tropical waters to identify the coral heads, as they adjust for the refraction of light and make cruising in shallow waters that much safer.

If one of the crew does become badly sunburned, firstly cool them off with lots of cold fresh water on the burned area. Then once dry, treat with one of the anti-burn sprays. If the patient is still in a distressed state, get them to a local doctor for expert attention as soon as possible.

Unwanted Visitors

The most frequent visitor and in many ways the most detested pest that can appear on board is the cockroach. These little creatures are everywhere in the tropics and Mediterranean and once on board are very difficult to remove. If ever there was justification in the saying prevention is better than cure, then this would clearly relate to the cockroach.

Sensible ways of keeping them at bay are, always to wash fruit and vegetables thoroughly before bringing them on board ship. As roaches lay their eggs in the folds of cardboard boxes, for the young to feed on the glue after hatching, don't carry the boxes past the gangway. It is also wise not to lay alongside for too long, as the little beggars will simply stroll on board when you are not looking. Try to moor bow or stern to, and if using a gangplank, keep this

raised an inch or two off the dock at all times.

If despite all the precautions you do get an infestation, what's to be done? Fumigation will rid the boat, but it is very inconvenient and extremely smelly, you will have to vacate the yacht for the day and the chances are that the whole procedure will need repeating, when the eggs hatch out a few days later.

Roach Hotels can be used; these are little boxes with bait and a sticky substance inside that attracts, and gums the little devils to the box, for disposal at a later date. Banging them on the head will work if you hit them hard enough, but it's difficult to catch them at the best of times, as they seem to have a very strong instinct for self preservation and soon learn to show themselves only when everyone on board is sleeping. It is a well established fact that cockroaches will eat virtually anything and thrive on such diets as cardboard and bilge litter quite happily. The problem is that they multiply rapidly, so from relatively modest beginnings, a yacht will soon have a major infestation on its hands unless war is waged on the pests, just as soon as they are discovered to be on board.

Boric acid is one way of poisoning them, and some cockroach baits have this as the main ingredient, which once digested apparently pierces the gut of the unfortunate creatures.

Another method I have heard of is to give them 'the pill', which, when crushed up into a powder is irresistible to them, when eaten it turns them pink and kills them in short order. Sprays can be purchased that slows them up a little, but as most of them live in inaccessible areas of the boat it's not really a long term cure. The best remedy is to keep them off the boat in the first place.

Pollution

It is a sad fact that some areas of the Mediterranean are heavily polluted, which to a large extent has been caused by the massive growth in tourism over the past twenty years or so. Unlike other oceans in the world it's renewal is slow, due to it being almost totally enclosed, with only the Strait of Gibraltar offering a current over its sill into the Atlantic. The change of water has been calculated at something in the region of one hundred and seventy years for complete renewal; so what pollution is put into the sea is likely to remain there for very many years to come.

The main problems come from industrial waste being pumped directly into the sea, tankers illegally washing out their tanks and large town and hotel complexes pouring raw sewerage and general litter into the sea around the coasts. The picture is not all gloomy of course, a great deal of coastline is unpolluted to any major degree and the water can be wonderfully clear and clean. Plastic unfortunately tends to float around a lot, much of it in sheet form, sometimes from the large plastic greenhouse growing areas that abound in southern Spain, but in general the picture is not too bad. The problem will come in a decade or two's time if the regulations on pollution are not more rigourously applied.

As yachtsmen we can do our bit for the environment by throwing nothing that cannot biodegrade quickly over the side. Also maintaining our diesel engines properly so that smoking is kept to a minimum and being careful about spillages of oils and fuel at all times. A positive move by some marinas is the availability of waste oil tanks, where the old engine oil can be tipped when replacing lubricating oil in our engines. Every little helps and if we are all aware of the problem, little by little we can help the environment recover, to the benefit of us all.

Tropical Waters

Cruising in tropical waters calls for a slightly different approach from the European or Mediterranean sailing grounds.

In the Caribbean it is largely eye-ball pilotage, since the conditions are normally excellent in terms of visibility and sometimes high landfalls can often be seen thirty or in exceptional conditions forty miles distant. Fog is almost unknown, but haze can restrict visibility to a marked degree at times. Care must be taken when sailing from island to island to allow for the current which is quite strong in some places; so what originally started out as a reach, could easily develop into a rail down beat if drift is not allowed for.

The most wonderful aspect of Caribbean sailing is the warm north-east trade winds, that blow the year round so reliably. They provide superb sailing and if cruising from the Windward to the Leeward Islands, most of the time a yacht will be on a reach at close to maximum speed. This exhilaration is enhanced by the large deep blue seas that tumble and roll across the Atlantic, they have a special quality that only seems to occur in this region – or perhaps it is just the magic of the Caribbean way of life casting its spell. The result is wonderful and surely this must be one of the finest cruising ground to be found anywhere for sailors who love the sea.

The downside of cruising this area must be the shallow waters of some coral islands where care in pilotage is a real necessity. Navigation is straightforward enough, but lights are not. Sensible yachtsmen don't rely on charted lights, for they can be very unreliable. It is best not to sail at night at all, as many islands and off-lying dangers are unlit and the currents can be unpredictable. Sailing from one island to another is usually a comfortable days sail, but it is important to arrive at the chosen destination when the sun is high, never, if at all possible, when it is between the anchorage and the yacht. An early start, and an early afternoon arrival, is the way in the Caribbean.

When approaching a shallow or coral

anchorage get someone as high as possible in the rigging to con the boat through the rock outcrops, particularly if it is a tricky approach.

Sometimes it is difficult to judge the precise depth of shallow water over a coral reef, but there are some rules that are worth remembering. Dark blue water is deep, pale blue or green is shallow water over sand and brown is coral. If the weather is overcast, the same general principles apply but the colours take on a more muted tone and look more brown than green. In these circumstances approach with extra caution and at a greatly reduced speed. Remember, coral is rock hard and can do a great deal of damage to a yacht.

Swimming & Snorkelling

The inviting warm and friendly waters of the Caribbean are renowned for their quality and clarity. Fish abound around the reefs and snorkelling can be a wonderful experience. As the waters are so warm, it is sometimes difficult to judge the time one has been swimming, so care must be taken to reduce the possibility of sunburn by using waterproof sun screens and some form of cover for the back.

If snorkelling in an area where there is a great deal of activity from other boats, such as water skiing, always have a small buoy with a little flag attached to a pole on a fathom length of light line tied to your leg. This will alert the other boats that someone is in the water, or maybe a few feet under it. It will save possible accidents and increase the enjoyment of the swimmer due to the knowledge that his position in the water is clearly marked.

Being bitten by fish is an understandable fear of many people – myself amongst them. The facts are that this is extremely rare in the Caribbean, but to be on the safe side a few simple rules should be understood. Sharks as a rule are at their most dangerous when in deep water. Close to

reefs there is an abundance of food so presumably they are not hungry enough to attack humans. Don't swim at night, as fish tend to feed at this time when there is more aggressive activity. If when swimming you do see a shark, don't panic! Turn and face it, trying to manoeuvre so that your back is towards the boat or the shore, then quietly and gently, and with the minimum of splashing paddle to safety. This, I emphasise is what I have been told by experts in these matters. It thankfully has not come from personal experience, if it had, I don't think I would ever venture into the sea again.

Other dangers that are more likely, come from such things as sea urchins whose spines are poisonous and if trodden on can cause terrible pain. The best treatment is to completely remove the broken off spines from the foot and bath with warm antiseptic water. If the pain persists get expert medical help. The stone fish is a serious hazard. Its habit is to lie on the bottom of the sea partially covered with mud or sand with just its spine sticking up, waiting for someone to tread on it. It has very poisonous venom that can seriously damage ones enjoyment of life, and has been known to be fatal. Treatment is to clean and wash the wound with hot antiseptic water, then get ashore for medical treatment with an anti-venom serum. The Caribbean equivalents are the scorpion or barb fish which are equally unpleasant creatures and to be avoided.

Sea snakes are fairly common and have poisonous bites the same as their cousins from dry land. These snakes live almost all their lives in the sea, so its reasonable that at times, presumably during their mating cycle, that rafts of the creatures can be seen together. I am reliably informed that they are not aggressive, but can be quite inquisitive, and gather around a swimmer. If bitten, (which is rare) get ashore as quickly as possible to seek medical help. If none is available, bandage the area and get the victim to lie as quietly as possible to localize the poison. Contact a doctor or professional via the radio and keep the victims air passages clear, give mouth to mouth if necessary.

Coral cuts and abrasions are fairly common and should be treated with an antiseptic wash then dusted with antiseptic powder or ointment. The wound should then be kept dry until healed.

The Caribbean is a paradise, but care and common sense should prevail. We always use hard soled beach shoes when walking on coral or rocky beaches; sometimes leaving them on for swimming if the water is shallow over a coral or rocky bottom.

Care of the Hull – Underwater

Keeping the hull clean of speed-sapping slime and weed in warm waters can, at times, be something of a problem. Marine growth in certain places is quite severe, whilst conversely, other areas enjoy less fouling.

One big consolation however, is that hull cleaning is reasonably easy in clear warm waters, for at least you can see what you are trying to clean off! We have found that a half an hour of manoeuvring, wearing mask and flippers and armed with a long handled deck-brush will see most of the hull wiped clean. If this chore is carried out a couple of times during the sailing season, there is no real necessity to have the hull pressure washed when hauled out for winter lay up.

If swimming around your yacht doesn't appeal, there is a very good hull cleaning device that enables the crew to clean the underneath of the hull whilst standing on a pontoon. It is called the Hull-Maidä. This is basically a wide brush with a long adjustable handle to reach the bottom of the keel. Air pressure keeps the device buoyant and pushes it against the hull to help removal of the fouling.

Choosing an anti-foul paint is not a problem either, as there are usually some excellent local paints available. If in any doubt, ask the local boatman or fisherman what they use, for it is usually reasonably priced – and red!

For several seasons now we have used an American produced anti-foul with excellent results. It is Awlgrip Gold Seal™, which will give protection for up to two seasons if two to three coats are applied during fit out. Normally, as we wipe off the light fouling afloat once or twice during the summer, the boat comes out of the water as clean as a whistle in the autumn.

Propellers

I am not in favour of using anti-fouling on props, particularly on sailing yachts. Far better to keep them really smooth, by rubbing them down with very fine wet and dry paper, then finishing off with metal polish. This will produce a highly polished slippery surface that will maximize the efficiency of the propeller and cut fuel consumption as well. It is worthwhile to check and clean up the propeller tips every time the yacht is slipped. If there are any minor nicks in the blade tips, gently smooth them out with a fine toothed file. If you find any blade has been bent, the prop should be removed and sent to a specialist for repair and re-balancing.

Some sort of protection for the actual propeller is desirable for warm water sailing. The lobster pot is in flourishing health and is everywhere. It seems that fishing from small craft is a way of life for almost the entire population of these areas, for they litter the inshore waters with their pots and floats. We have been unfortunate enough to foul our propeller on three separate occasions, in spite of taking great care in threading and weaving our way around headlands, (areas much loved by the pot people). In every case the propeller has been saved from damage by the little rope cutter

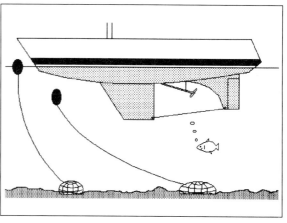

A clean burnished propeller will give better performance than one covered with anti-foul.

A useful method of preventing stray lines fouling the propeller is to run a wire between the aft end of the keel and the leading edge of the skeg.

we fitted to the shaft in England. This device comprises several hard cutting teeth made from stainless steel, which revolve with the shaft, and snip off any rope or line that attempts to wind its way round the propeller. All in all they're efficient, they happily require little maintenance, but give a great deal of peace of mind, particularly when motoring inshore.

If working below the surface is something that appeals, there is a small diving apparatus sold that can be used by unqualified persons. It is called a 'Sweeba™' and has been designed for safe diving up to two metres in depth.

With such equipment, servicing the underwater hull would be a relatively simpler matter and I am sure it will prove a boon to many yachtsmen. Speaking personally, I prefer to stay on the surface if at all possible – so its appeal to me is somewhat limited.

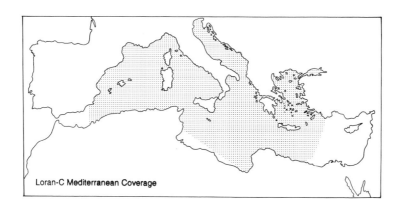

Loran-C Mediterranean Coverage

Chapter 17

In Search of the Ideal Cruising Yacht

Of course, the perfect cruising yacht has never been built. There are however some boats that are clearly more suited to this role than others. Many of the most desirable cruising attributes in yacht design are well known, but this changes, new materials are found and new techniques are discovered.

I thought it would be an interesting exercise to examine three specialised ocean cruising designs that are currently available on the market.

They have been chosen as good representatives of their type, for they embody most of the important attributes I consider essential for safe, fast offshore cruising. They are however, very different in character and perhaps this reflects just how far modern design has progressed in the past twenty five years or so, for in spite of

their differences they are all outstanding yachts in their own right.

When evaluating them, they were put through a series of sailing trials, plus a critical appraisal of build-quality, design and equipment levels. It is satisfying to record that common sense has prevailed and they reflect in large measure a wholesome regard for the true requirements of offshore cruising.

None slavishly follow the fashionable trend of the IOR(International Offshore Rule – a racing handicap system) influenced hulls and in my opinion are the better for it. There is little doubt they will all be good sea boats with the potential of achieving high average speeds safely without stressing their crews too much and this must surely be the very essence of what a good cruising yacht should be.

Tandem 38

A semi custom design built of wood

On first sight the Tandem 38 *Fondue* looks to be a fairly conventional masthead sloop. Slightly angular in appearance, with a chopped-off stern to produce a scoop for easy boarding. She has a no-nonsense, purposeful air that is enhanced by her high quality deck hardware and oversize rigging, all of which gives a clue to the thought processes that have gone into her design and construction.

Born from the mind of an experienced owner, who wanted a yacht capable of swift, long distance cruising in comfort, the concept centres around the revolutionary tandem wing-keel of Warwick Collins, who also designed the hull. I have been interested in the tandem keel for some time

and was keen to find out just how good it was and how suitable it might be for serious ocean cruising.

To date some hundred tandem keels have been fitted to a variety of yachts from the little twenty five foot ULDB *Flying Boat* to the superb eighty foot ocean cruiser *Ocean Leopard*. (*Ocean Leopard* with full cruising equipment aboard, smashed the monohull Round the Island Race record in 1991, leaving her conventionally keeled sistership over thirty minutes behind.) It would therefore be fascinating to learn how *Fondue* handled, for the keel and hull have been designed to compliment each other.

The deck and interior design was a product of the strict brief laid down by the owner, as was the LOA/beam ratio which has led to a fairly narrow boat by today's standards.

On going aboard one is immediately struck by her functional quality and her stiffness. The cockpit is spacious and well protected by the generous spray-hood that shelters the companion-way. Sensibly there are full width drain scuppers built through the transom. A shallow bridge-deck leads down to the wide steps into the accommodation. The pushpit has been designed to accommodate an instrument gallery. The radar is mounted on its own strut clamped independently to the transom.

The mast is a two spreader Isomat™ spar with a spinnaker pole attached to its forward side. A separate trysail track is positioned independently on the port side of the main track. Runners have been included, but I understand are rarely used. All lines are led aft to the cockpit under a garage via a battery of stoppers, then to fabric pouches fastened to the coachroof bulkhead. The cockpit has a row of shallow coaming lockers to take winch handles and the like. The helmsman sits on a humped seat behind a small stainless steel wheel; the position offers good command over the ship and visibility of the pedestal mounted compass. Genoa and spinnaker winches are close to hand.

The mainsail horse is positioned on its own full-width track forward of the spray-hood and the sheet is led aft to a coachroof mounted winch. This is an excellent system that works well in practice. It helps to keep the cockpit clear of extra lines, without reducing efficiency in any way.

Anchor handling is conventional with a deck-mounted winch. The roller fairleads are large and well positioned clear of the stem. A separate detachable three-quarter length forestay is used for storm and working jibs. The genoa is full sized on a roller headstay, with the roller drum positioned below deck level in an enclosed well. A fully battened mainsail is fitted, with lazy jacks and the boom has a solid kicker for control.

There are watertight bulkheads fitted fore and aft and all lockers are watertight and lockable. In the unlikely event of a localised impact causing a rupture of the skin, the incoming water would be contained.

Below, one is immediately attracted to the airy open feel of this yacht. The finish is a mixture of traditional varnish and white paint. The main saloon bulkhead that separates the accommodation is well cut back to the yacht sides, which helps the air of spaciousness. The owner specified athwartships position for the chart table. It is certainly spacious, with excellent storage for charts, books and instruments; there is a narrow 'seat' to lean on. (I feel however, that it would not be as convenient or comfortable on a long cruise as a fore and aft position, particularly beating to windward. However, the owner says it's 'a great success', so perhaps my assessment is incorrect.)

The galley is very well designed, but lacked a secure 'bum strap'. The floor is varnished teak and will soon become slippery. I feel some attention to this point would be desirable before undertaking serious offshore work.

Aft of the galley area is a small double cabin. The bunk has a split cushion to accommodate a lee cloth. There is good headroom here and it is a practical, if rather cosy, sea cabin for two people.

On the starboard side aft, there is a roomy head compartment with built in wet clothes locker. This is warmed by the engine heated calorifier, which is a nice touch. Main saloon storage was rather limited as the water tanks were positioned under the bunks.

The forward cabin was, I felt, a little cramped due to an additional heads compartment being positioned in the area. This is rather ambitious for a thirty eight foot, fairly narrow beamed yacht and I felt the space could have been better utilized for

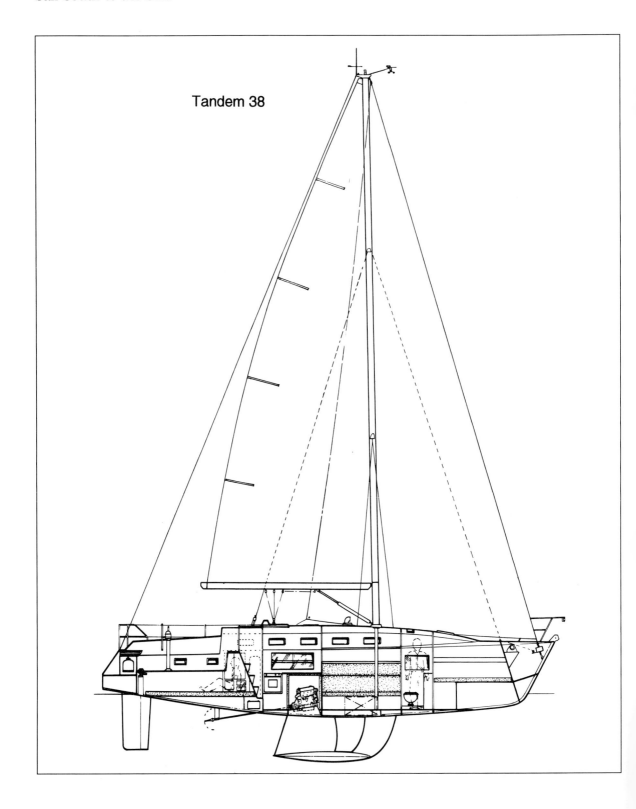

Tandem 38

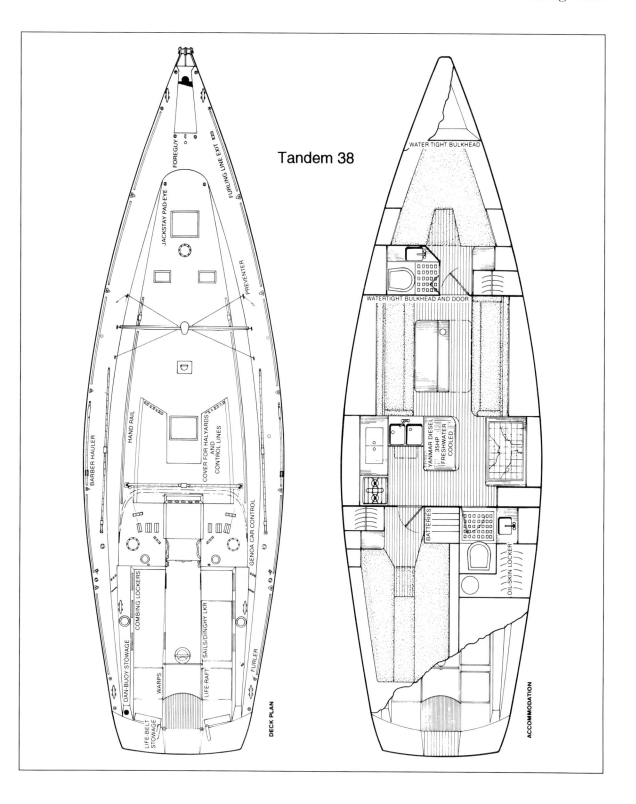

Tandem 38

DECK PLAN

FOREGUY

FURLING LINE EXIT

JACKSTAY PAD-EYE

PREVENTER

HAND RAIL

BARBER HAULER

COVER FOR HALYARDS AND CONTROL LINES

GENOA CAR CONTROL

COMBING LOCKERS

DAN-BUOY STOWAGE

WARPS

LIFE-RAFT

SAILS/DINGHY LKR

FURLER

LIFE-BELT STOWAGE

ACCOMMODATION

WATER TIGHT BULKHEAD

WATERTIGHT BULKHEAD AND DOOR

YANMAR DIESEL 35HP FRESHWATER COOLED

BATTERIES

OIL-SKIN LOCKER

139

Sparkle, the second of the Tandem 38 class has teak decks and a cruiser oriented fit-out.

Her cosy practical saloon typifies the thought that has gone into this offshore design.

Effortless sailing in Force 5.

Fondue shows her clean no-nonsense lines.

Note control lines led under removable garage.

additional stowage, with perhaps an addition of a small hand basin incorporated for the occupants.

The engine is housed in its own centrally positioned box in the saloon. There is a work top for the galley incorporated above but it does not intrude in any way. Whilst running, the engine was extremely quiet, due I was informed, to the high level of insulation of the box surround and flexible engine mounts.

The cooker is a two burner gas model with oven and grill, gimbaled and with a substantial crash bar. Gas stowage was located in the cockpit. The bottles had their own compartment under the helmsmen's humped seat, with overside drain incorporated.

The design brief for comfortable living for two couples is clearly met and she should be practical to live on for extended periods.

Sea Trials

I was lucky, in that the day selected to sail her turned out to be fairly windy, ideal for testing her role as a blue-water cruiser. A brisk Force 5 was forecast and I was particularly keen to see how she handled in the short choppy waters of the Solent in wind against tide conditions.

Her general handling was excellent under power. She backed out of her berth, immediately coming under command just as soon as way was established. Her turning circle was small, so handling in confined waters was no problem.

The fresh wind necessitated one reef in the main and six rolls in the genoa. With twenty two knots True over the deck she punched to weather at about six and a half knots. Her stiffness was most impressive and she showed no tendency to gripe to windward in the harder gusts. She had just about the right amount of feel on her helm, which remained light right through the test. An excellent attribute for long distance, short-handed sailing where an autopilot or windvane would be used almost continuously.

I ran her off – first on a beam reach and then unbalanced her with just the mainsail pulling. She did not gripe but simply accelerated smoothly, leaving a flat wash in her wake. Again helm movements were minimal and she showed little tendency to roll – she was very soft in her motion. We tacked all over the western Solent to find the roughest water, where her stiffness and stability were most impressive under these difficult short seas, boding well for the owner's plans for blue-water cruising in the coming years.

Conclusion:

As a cruising design this yacht is hard to fault. She is fast, stable and weatherly. Her motion in a seaway breeds confidence. She goes about her work in a manner of a much larger boat. She appears not to slam when working to windward, which is probably the result of her fairly deep and vee'd forefoot design. The flat run aft gives a soft ride with just a modest quarter wave even when footing fast off the wind.

Her construction is very strong and yet light; she held her way well through steep waves, much in the manner of a traditional heavy displacement design. This must be a function of the clever tandem keel. I was told by the owner, the rougher it gets the more impressive the keel behaved.

On a personal level, I felt some aspects of the design could be improved. The extra head was to my mind rather a waste of space in a thirty eight foot boat. Overall storage was rather limited and should be improved if long distance cruising is to be contemplated. Ventilation was designed for northern European waters; the aft cabin in particular would be very stuffy in a hot climate. What she needs is an extra hatch plus four large dorade vents. A simple cold water shower should be fitted on the transom for washing after bathing and a ladder incorporated in the stern for easy boarding and possible emergency rescue.

She is however a brilliant design with many of the most desirable features necessary in a blue-water cruiser brought together in one yacht.

I was particularly impressed by this yacht's high stability, which in my opinion is an essential quality all ocean cruising yachts should possess. The design has been analysed by the Wolfson Research Unit and the results for the angle of vanishing stability of the Tandem 38 in cruising trim is over 160°. The tandem keel's low centre of gravity must play an important part here. Another impressive factor regarding this yacht was its very tight turning circle. With a conventional fin keel it stalls at approximately 12° of leeway causing disruption of flow and increase in drag. The tandem keel does not stall until well over 30°, and in consequence looses less speed and continues to grip without drifting outwards. This is all the more remarkable when one considers this boats excellent straight line 'tracking' ability, a combination of attributes that must be almost unique.

Cost:

To build another Tandem 38 to *Fondue's* high specification today (1992) would cost approximately £150,000.

Tandem 38
LOA 38ft BEAM 11ft 4in
DRAUGHT 5ft 4in
DESIGNER: Warwick Collins, 23 Kingston Park, Lymington, Hants, England, SO41 8ES
BUILDER: Various
HULL CONSTRUCTION: West Epoxy System™. 1 skin.
15mm red cedar, edge dowelled and stapled to laminated khaya frames; 500mm centres plus 3 additional skins laid fore and aft. Finished with full layer of 290gsm GRP mat.
DECK CONSTRUCTION: 2 – 4 layers of West Epoxy 6mm plywood, painted.
KEEL: 'Tandem Wing' by Warwick Collins – iron casting.
RUDDER: Wood coated in GRP mat. Full length skeg.
DISPLACEMENT: In full cruising trim approx 7.2 tonnes.
ENGINE: Perkins Prima 29hp diesel on flexible mountings. Shaft on P-bracket.
PROPELLER: Two blade, folding.
SPEED UNDER ENGINE: Maximum in smooth water – 6.9 knots; Cruising (2000 revs) – 5.5 knots.
WATER: 90 imp galls in two tanks.
BALLAST RATIO: 48%
HEADROOM: 6ft plus.

Vancouver 36'

A heavy displacement GRP Cutter

The Vancouver 36' is the flagship of the Vancouver range of yachts built by Northshore Yachts of Chichester.

This well respected range of chunky cruisers of 28ft, 32ft and 34ft have in the past been designed by Robert Harris and some remarkable voyages have been made in many of them over the years. For the

design of the 36', Northshore went to Tony Taylor, who for more than 20 years was with Camper and Nicholsons of Gosport, England – and what an excellent job he has made of it. Northshore Yachts mould their own boats so quality control is strict. They have also developed a unique laminating system called Nordseal™ which they claim helps protect the hull from osmosis attack. The addition of a coating of Cystic Copperclad 70 PA™ gives a measure of anti-foul protection for up to 15 years.

I inspected some of their boats inbuild and was most impressed with the general hull lay-up and solid construction of rudder and keel members. The hull is moulded in two parts and joined fore and aft down the middle. The ballast is of lead and is encapsulated low down in the keel, with the draught of five foot seven the boat will be stiff.

On climbing aboard I was immediately taken with the solidity of the boat and size of the fittings and rigging. Here was a little ship that had a purposeful air about her.

On Deck

The Kemp mast supported a cutter rig. On this particular boat the owner had specified an in-mast furling mainsail system. All sail controls lead back to the cockpit via a battery of clutches on the coachroof aft, under the generous sized hood. There was heavy Treadmaster material on the decks and an excellent non-slip finish moulded into the coachroof. Footing all round felt secure as it should on a boat designed for offshore work. Stanchions were a good height with double lifelines and gates port and starboard. I particularly liked the substantial pushpit with the bathing ladder integrated as part of the design. The stemhead fitting was suitably strong with twin rollers, one of which handled the forty five pound CQR anchor when shipped. This is backed up with thirty fathoms of three-eighth calibrated chain. There is a hand operated windlass on the foredeck, which is now a standard fitting, plus a large twin horned bollard for mooring. The stern houses two ten inch mooring cleats and two others were fitted amidships for springs.

Sails and Spares

All sails are by Arun. The standard inventory includes a Yankee on a Kemp furler, plus a staysail and mainsail. These were made to the heavy cruiser specification with triple stitching.

The mast is by Kemp, and although cutter rigged, the set up is considered stiff enough to dispense with runners. There are, however, fixed intermediate backstays to support the inner forestay.

The spreaders were of a good length so helping to reduce compression loads on the mast section. The boom is fitted with a solid kicker.

Sail controls were convenient, with the main track mounted on the coachroof forward of the cockpit hood. Lines were then led directly back to one of the six Lewmar self- tailing winches.

Cockpit and Steering

The cockpit is first class, large but deep with high, correctly angled coamings with nicely rounded corners to snug down into. Drainage is fair. The big folding spray-hood protects the forward part of the cockpit well and visibility is excellent due to large clear plastic windows let into its forward and side faces. A humped helmsman's seat is provided with life-raft storage under. Engine controls and compass are mounted on the binnacle. Steering is a Whitlock Cobra R system with direct linkage to the rudder. An emergency tiller is supplied. The thirty inch wheel is hide covered. As the forward part of the cockpit is wide, a removable teak footbar is fitted for crew convenience, this is a good safety feature. A folding cockpit table can be made to fit onto the binnacle if required. Teak gratings are supplied for the sole and teak laid decking covers the cockpit seat areas.

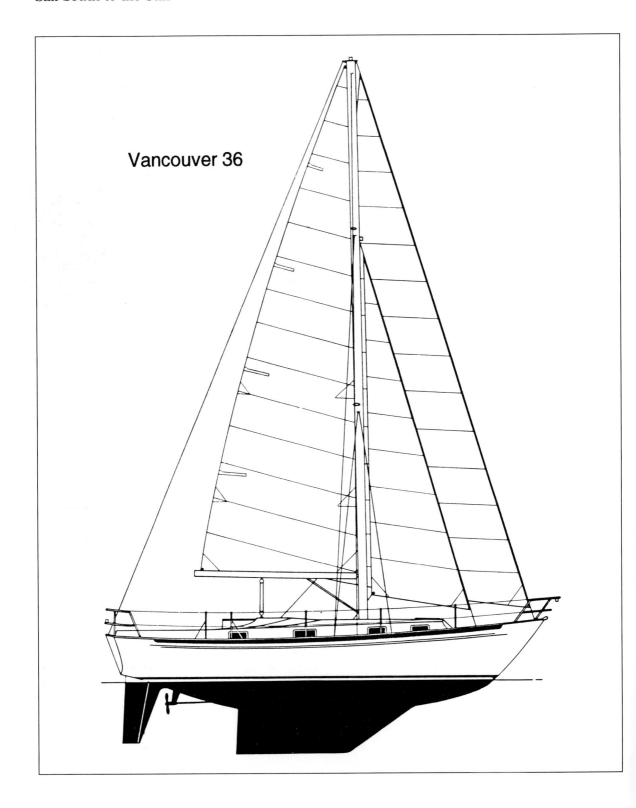

Vancouver 36

Vancouver 36

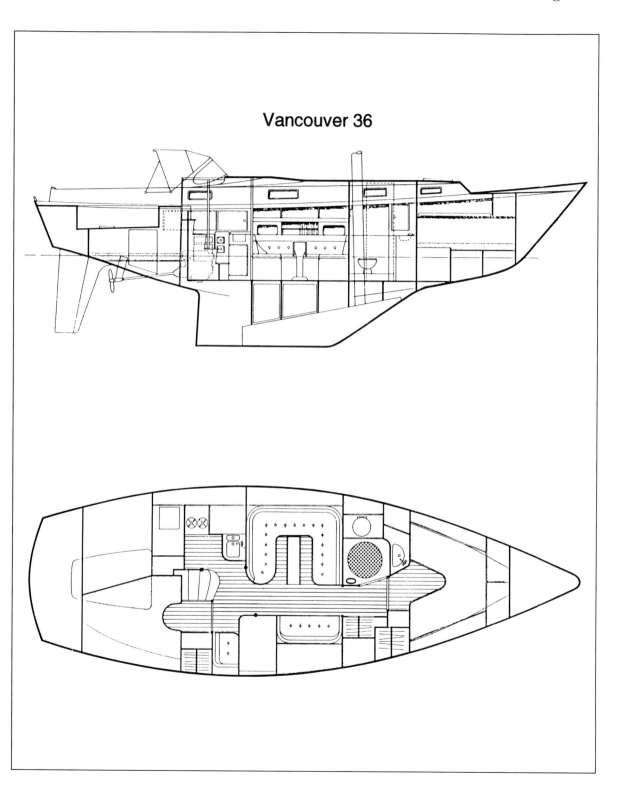

Vancouver 36

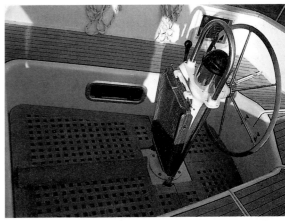

Note foot bar and folding table in large cockpit.

The well designed chart table and navigatorium

The large spray hood is an excellent feature.

Well equipped galley with plenty of handholds.

Below Decks

Stepping over a low bridge-deck and descending below is simple, due to the wide companion-way with secure handholds. The footing is covered with a non-slip material that seemed most effective.

The general feeling was of light and airiness due to the wide saloon with well cut back bulkheads. To port is an excellent U shaped galley. Work surfaces are adequate and the twin sinks (one large and one small) are sensibly deep. Pressure hot and cold water is piped here and to the heads.

There is a large ice box that can be converted to a refrigerator. Crockery stowage is well designed and secure. Good ventilation of the galley is ensured by a separate opening hatch plus an electric two way ventilation fan.

I was less happy with the safety aspects regarding the cook keeping his feet in a rough sea. There was no galley strap fitted and the floor was varnished teak. These are very important considerations which could be put right very easily with little additional expense. The cooker fitted to this particular yacht was a good quality model with twin burners, oven and grill with 'fail-safe' on all burners. It was gimbaled and a substantial crash bar fitted. General stowage was excellent with cupboards, lockers and drawers for cutlery etc.

The large forward facing chart table is to starboard. There is reasonable stowage for charts in the locker under the lid. The seat is firm but has no lateral support for the navigator when heeled on the starboard tack. Instrument bulkheads are generous but I felt the stowage for books a little narrow and not very practical. There is a good tray for odds and ends however.

Saloon

The saloon is most comfortable with a U shaped dinette to port surrounding a solid oblong table fitted with deep fiddles in its centre section. An excellent fold-out pilot berth is provided over the dinette plus another fixed berth with trotter box on the starboard side outboard of the short settee.

An excellent feature is a roomy oilskin locker behind the galley at the foot of the companion-way. It can take several suits of oilskins and is heated by the engine which is housed under the companion-steps. There is plenty of usable stowage, every nook and cranny has a purpose and used to good effect.

Sleeping Cabins

The aft double cabin is of good usable size for a thirty six foot boat. Standing headroom is about six foot six and there is enough room to dress without too much effort. A wardrobe is fitted, together with a sizeable vanity table and mirror. The berth cushions are split down the centre and lee cloths are provided, so this is a practical one or at a pinch, two berth sea cabin.

The forward cabin is fitted with a conventional double V berth arrangement, with a wardrobe to starboard. Good underbunk stowage is in GRP bins plus lockers and shelves. There is also a vanity unit and mirror fitted here. Both cabins have large ventilation hatches.

Heads

Sensibly, in a boat of this size, there is only one head compartment. It is however large and well fitted out with shower stall with cover for the loo seat, a single basin with hot and cold water and a whole range of lockers for toiletries etc.

Waste, drains through a teak grating and is moved overboard by an electric pump. The head compartment has good handholds and in general is an excellent feature of the boat.

Machinery

The particular yacht I tested had a Volvo 2003T 3 cylinder engine installed. This is a turbo-charged model that develops 43hp at 3200 rpm. It drives a 20in fixed three bladed prop. The engine is fresh water cooled and heats the domestic hot water via a calorifier. The engine room is ventilated with a forced draught electric fan unit.

All seacocks were accessible and hoses are double clipped. The electrics panel houses a battery of switches all protected by circuit breakers. It is hinged for access and maintenance.

Sea Trials

I was particularly fortunate in selecting a day of fresh winds for the sea trial. Force 5-6 was forecast and if anything this turned out to be underestimated. Perfect conditions for putting a boat designed for offshore cruising through her paces.

With a reefed Yankee plus full staysail and the equivalent of a double reef in the mainsail we beat against a young flood in Chichester harbour to approach the bar. As we arrived early at Hayling Island there was insufficient water, so we used the waiting time to tack and reach about to judge her agility under sail in a confined area under fresh conditions.

She was stiff in the gusts, showing little tendency to gripe to windward when hard pressed. The helm was a little heavy however, but I was assured that this was due to the autopilot being connected. Unfortunately I did not have the opportunity to judge if this was so, as it was not convenient to unship it in the conditions that prevailed. She was well balanced and tracked well in the smooth water of the harbour.

At half flood we put her to windward and crossed the bar to sea. Conditions were fairly rough with a 25 knot wind against tide, but she made mincemeat of the short seas and crunched her way to windward displaying the merits of heavy displacement in a boat of this size. She remained fairly dry with only heavy spray coming over the deck. Her stiffness was most impressive, sailing more like a larger boat than one of only twenty eight foot waterline.

Her pointing ability was difficult to judge as the steep seas ensured we had to sail her a little free, also the mainsail in its deep reefed position seemed to be short on drive. Certainly she was well mannered, running off down wind with the sheets pinned in, she showed no tendency to stall out or broach; although to be fair, she was snugged down well and not being driven hard.

All in all she showed a good turn of speed and was easy to handle. She was steady on the helm with no vices that I could judge, in the relatively short time we had available to us. I was less than enamoured with the in-mast reefing system, it seemed to have quite a lot of friction and could not be reefed or indeed rolled up fully, until brought into the wind. I believe this boat should have a fully battened mainsail, and I am not altogether convinced about the cutter rig on a boat of this size either. It seemed to break up the power of the rig too much and adds complication where none need exist. A sloop foretriangle would develop more power and on a thirty six foot yacht with the stability of this Vancouver, handling should not be a problem. Her directional tracking was good without being outstanding. For stiffness and balance under fresh conditions she was first class and I would have no worries about taking her offshore.

Under power she handled practicably although we were not involved in manoeuvering in a confined marina environment, but I was assured she behaved well when going astern. The engine felt a little harsh and seemed not to be producing the power in terms of thrust that I would have expected from 43 hp. It could be she was slightly under propped or, as it was close to the end of the season her bottom may have been a little foul.

Conclusions

All in all this is a superb conventional yacht that meets her design parameters for a serious offshore cruiser splendidly. Beautifully and strongly built she is a credit to her builders. The care and attention the designer has taken over all those little details, make her stand out from her competition. If she were mine I would change the rig to a sloop, give her some light weather sails, make one or two minor modifications, then head for distant horizons with complete confidence. She represents excellent value for the serious yachtsman.

Cost: with comprehensive standard specification is £99,500 (1992).

Vancouver 36

LOA 36ft
LWL 28ft 6in
Beam 12ft
Ballast Weight 8,198lbs
Displacement 20,494lbs
Draught 5ft 7in

Sail Areas
Mainsail 290 sq ft
Yankee 364 sq ft
Staysail 113 sq ft

Northshore Yachts Limited, Itchenore, Chichester, West Sussex. PO20 7AY England. Telephone (0243) 512611 Fax (0243) 511473

Premier 45

A heavy displacement, ocean cruising cutter built of steel.

Premier Yachts of Emsworth have set out to produce a range of high quality custom yachts specifically for ocean cruising.

In their dedication to the task they have drawn together a team of experienced people headed by Chris Kidd, Bill Dixon the designer and builder Edward Dridge.

The current range includes 41ft, 45ft 50ft and 55ft yachts. All are built of steel to a very high specification that includes teak decks and a comprehensive inventory of quality equipment.

I tested the first yacht to be launched, *Maid of Steel* which is the 45ft version with centre cockpit and cutter rig.

The yacht has a pleasing modern appearance with highish top sides and nicely tucked up counter stern. The coachroof detail is well proportioned and not too obtrusive. Although the toughened glass windows are large they can be protected at sea with removable storm boards if conditions demand it.

The underwater profile shows a slippery shape, the keel is reasonably long for good lateral resistance. The rudder, set well aft is deeply skegged and of generous area. The forefoot is quite deep and well vee'd, so she should not slam to windward in a blow.

The general impression on climbing aboard is one of strength, coupled with superb quality of equipment and fit out. The teak decks were beautifully laid with nice attention to detail. The cockpit was well protected and deep with a battery of self-tailing Anderson stainless steel winches. All sail control lines are led aft to a series of clutches.

Construction

The hull is built of 4mm steel on 44mm x 6mm T sections at 450mm centres. Deck is also 4mm steel on stout 70mm x 8mm beams, overlaid with 12mm ply and faced with 12mm Burma teak, so there is little doubt that she is strongly built.

Being somewhat concerned with the problems of corrosion with steel I was emphatically assured that 'corrosion is now a thing of the past if modern materials and techniques are correctly applied'.

All Premier yachts undergo a comprehensive fit out. Firstly all the steel surfaces, inside and out, are grit blasted, then immediately sprayed with molten zinc. The next stage is to epoxy seal and primer paint, five coats in the bilge area, plus two coats of two-part polyurathane paint on the topsides. Inside the hull above the bilge line is treated with wax then flame retardant polystyrene sheet. 1.5in thickness is applied to provide thermal insulation and minimise condensation.

The hull is fitted with the CAPAC current cathodic protection system, similar to the ones used on the steel oil rigs in the North Sea.

I was reassured to see that all seacocks were of the Lloyd's approved non-corrosive polypropylene type and that all were double clipped. Ballast is sealed within its own cavity in the keel and all voids are filled with oil to prevent corrosion of the inside plating.

Engine

A Perkins M90 normally aspirated 82hp unit on flexible mounting is fitted with a Hurth gear box. The propeller is a three blade feathering maxprop, protected with a rope cutter.

The engine compartment is well insulated and roomy. There is excellent access for routine maintenance and enough space to fit a generator if required. Fuel tankage is also built into the hull with a provision for approximately 120 gallons.

A deep sea seal is fitted as standard to the shaft.

Plumbing

Fresh water storage is not over generous for a serious long distance cruising boat of this size. There are two 120 gallon stainless steel tanks under the saloon floor coupled together with change-over valves. I am told that extra tankage could be built in future yachts if specified at the planning stage.

Pressurised hot and cold water is plumbed into the galley (which also has emergency hand pumps) in both head and shower compartments.

Hot water is produced by engine or immersion heater fitted to the 10 gallon calorifier. Both head compartments are fitted with shower stalls. The soil water is discharged overboard by an electric pump.

Electricity

There is a 24 volt negative earth system fitted throughout the yacht. Domestic power is supplied by four 6 volt batteries totalling 390 amp hours. There is also dedicated engine starting battery of 95 amp hours.

A Mobitronic battery control system is fitted, as is a 40 amp shore battery charger. The batteries are housed in a ventilated box with hold down straps. I was pleased to see the comprehensive control board was fitted with contact breakers and the wiring looms were all clearly marked.

Gas

A ventilated stowage for up to four Camping Gaz bottles is positioned in the side deck. All connections are in flexible copper and there is a gas shut-off valve located next to the cooker.

Steering

Steering is by a 36in leather covered wheel, with cable and guard-rail to rudder head. An emergency tiller is supplied. Also standard equipment is an Autohelm ST 7000 autopilot and Sestral compass, neatly installed in the alloy pedestal, together with a holder for glasses.

Ground Tackle

As befits a yacht designed for faraway anchorages, the ground tackle is well thought out and substantial.

A 60lb Delta anchor with calibrated chain plus a 35lb CQR kedge with chain/nylon warp are supplied.

Handling such heavy equipment calls for good stemhead gear. This particular yacht had a well designed bow roller fitting in heavy stainless steel on a short bowsprit. Coupled to the chain was a Francis 1500 electric windlass mounted on the foredeck, this is also standard equipment.

Rig and Sails

Mast and booms are by John Powell, anodised then painted to customers colour choice.

Rigging is by Spencers of Cowes, the hydraulic backstay tensioner has remote control.

Main boom control by Lewmar with solid gas vang.

All sails by Gowan. Staysail and foresail both have Harken roller reefing equipment.

The main is full battened with jiffy reefing and all controls are led back to the cockpit.

A cruising chute with spi-squeezer completes the standard inventory.

Accommodation

Accommodation is left to the owner to specify, due to the rigidity of the hull construction and the lack of structural bulkheads almost any layout can be accommodated.

This particular yacht catered for two couples. There are two double self-contained cabins with adjoining heads and showers. In addition for passage making, two single berths in the walkway to the aft cabin. Both double bunks have been sensibly split down the centre and fitted with lee cloths. This will at least make the aft cabin suitable for deep-sea work.

From the cockpit there is direct access to the accommodation via rather narrow but secure companion-steps.

The main saloon is light and airy with large windows. To port a comfortable D shaped settee/dinette and starboard two comfortable fixed armchairs. There are plenty of lockers in this area of the boat. With a built in TV cabinet and drinks locker above.

The chart table is one step down on the starboard side aft. Although not overlarge, it is adequate and took a folded admiralty chart. There was also excellent fitted book storage, and plenty of room for instruments in front and to the right hand side of the navigator. The instrument bulkheads could be removed for access and maintenance. There was a small locker for chart instruments and odds and ends.

The chart table seat was comfortable but needed lateral support in the form of a retaining bar. A foot brace for rough weather would also not come amiss.

The Galley

This was very well planned, although for my taste rather far forward in the boat, being ahead of the saloon. It was however just the right size with plenty of putting down areas, two deep sinks and a high specification two burner gas stove with oven and grill. The gimbaling was correctly carried out with plenty of space away from the ships side for 'swing' in heavy conditions.

There was a front opening refrigerator fitted, but I understand that a top opening model can be specified. Plenty of storage was a feature with secure burstproof

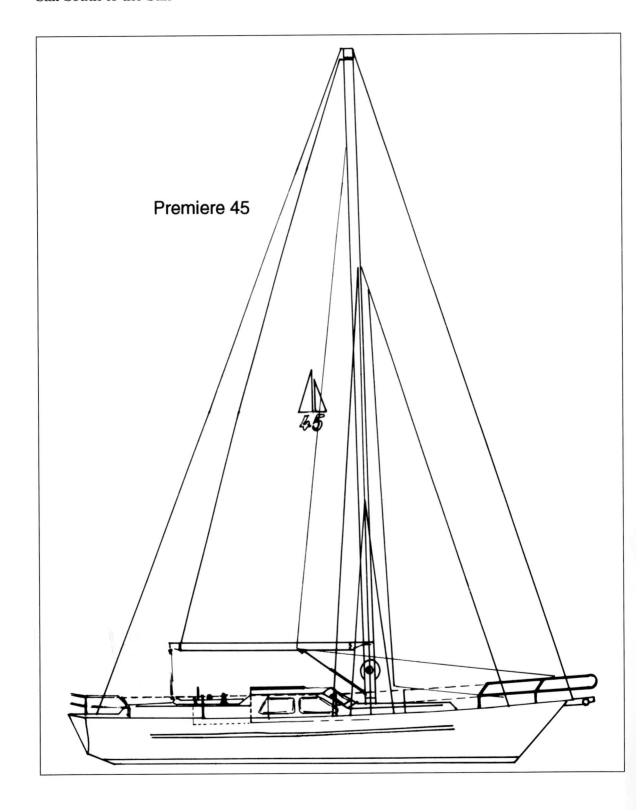

Premiere 45

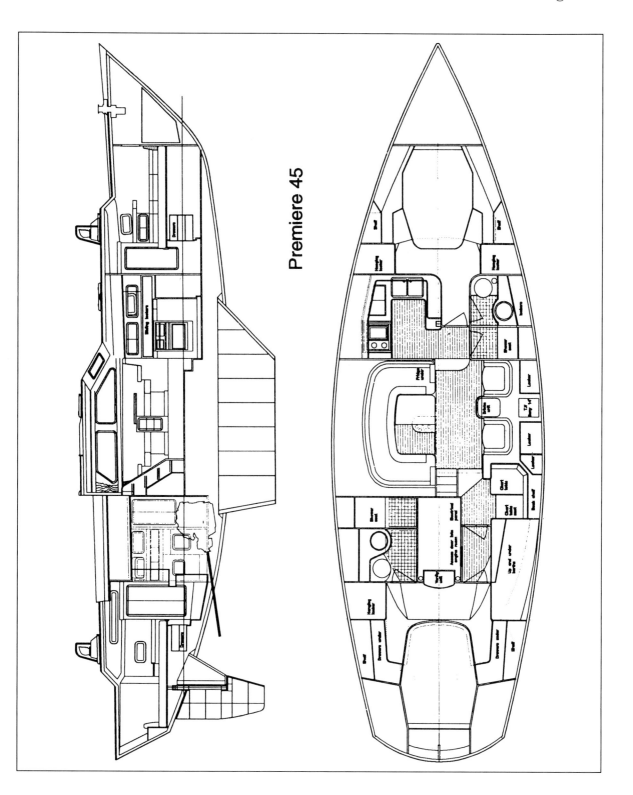

Premiere 45

Premier 45

The galley has extensive lockers,worktops and twin sinks.

The customized switchboard

Excellent instrument pod and engine controls in the well protected deep cockpit.

Mid ships spring cleat with roller fairlead.

latches. Working tops were covered with a thick marble-like material that would be easy to clean and remain cool in warm climates. A large hatch opened above the cooker to provide good circulation. There was also an opening topside portlight, a welcome extra ventilation hole, but to my mind not a sensible fitting for an ocean going cruiser.

The sole was coated with non-slip varnish, but although it looked smart, it would be lethal when wet. Some sort of real non-slip finish or battens screwed to the sole would be desirable before going to sea.

General Finish

The interior cabinet work was solid teak or teak faced ply. Some doors had an attractive ash in-lay decoration. There is I understand an option on the timbers used and owners can specify a mixture of teak, mahogany, holly or ash. The ceiling head panels were covered in a suede material held in place with teak battens. No fixings were on the surface giving a particularly high quality feel to the interior.

Deck Gear

All spars were to a high specification. A full length stainless steel pulpit and pushpit were fitted with sensible 30in high stanchions and double lifelines. Four dorade ventilators with stainless steel vents were positioned to give excellent ventilation under most conditions. There was a large hatch fitted over each cabin, and head compartment. Massive stainless steel cleats were positioned on the foredeck, afterdeck, and mid-beam for springs.

Sea Trials

The day selected to trial the Premium 45 turned out to be clear and cold, typical high pressure weather for late November. The wind was light, which although not ideal for judging her heavy weather capabilities would nevertheless be illuminating. Blue- water cruising is by no means made up of endless hard weather and gales, if anything the reverse is true. It is, in my judgement, very important for good noon to noon runs offshore, for a yacht to have good light weather performance. The trials therefore, would be a useful indicator to her capabilities as a sailing yacht in light to moderate conditions.

She is fitted with a bow thruster and powerful Perkins diesel, so extracting ourselves from the rather cramped marina berth was ridiculously easy. Once underway I put her through some basic manoeuvres under engine and she proved to be accurate in her steering. Her large rudder gave very direct control and she made a crash stop in one boats length from a full seven knots ahead. Her turning circle was compact, about one and a half lengths hard over. Manoeuvering astern was really superb, immediately she started to get under way, she was under full control. The medium to heavy displacement steel hull gave her a powerful motion, with little pitching in the seas we met in the Solent.

Hoisting full sail was simple, for all the control lines were led aft to the cockpit and we soon had her making to windward in about ten knots of breeze. Being very cold I estimated that the air would be denser than typical summer winds, so the pressure on the sails would correspond to perhaps thirteen knots in summer.

Her steering was comfortably light, bearing in mind the large rudder, and she footed ahead at around 5.5 knots with just a touch on the wheel and 8°/10° of heel. She felt stiff and responsive, her directional stability was outstanding; the wheel could be left for minutes at a time without attention and she only started to waver slightly if the wind altered direction or strength. The autopilot will have an easy life on this boat.

The cutter rig seemed powerful enough but the back-wind from the staysail seemed to interfere with the leech of the genoa, so we rolled the small sail up completely, with no apparent reduction in speed.

Once or twice the wind roused itself to twelve or thirteen knots and the boat immediately reacted by smoothly increasing speed. I was particularly impressed when going below by the uncanny silence of this yacht, there was not a creak or groan to be heard, only the slight rustling of the bow wave. Obviously the comprehensive internal insulation on the steel hull contributed to this silence but credit must also be given to the well fitting joinery.

Later, as we reluctantly turned for home, the cruising chute was hoisted to make the most of the breeze. It produced seven knots on occasion which was very creditable for this large, heavy yacht under these light conditions. Motoring back to our berth the boat managed seven knots at a peaceful 1800 revs, with a maximum of around seven and three-quarter knots. Berthing again was simplicity itself but we were pleased to get out of the cold and into the cosy warm of the main cabin. Who said central heating on boats was a luxury!

Conclusion

It would be difficult to imagine how this boat could be improved upon for ocean cruising, apart from detail accommodation layouts, which is to the choice of the individual owner. She proved to be fast and responsive with impressive directional stability. It is true, that conditions were light, but I see no reason to disbelieve the builders claim that in heavy weather she proved equally impressive.

Her sail handling systems have been carefully worked out. The single line reefing system led back to the cockpit and with twin foresail reefing, operated via Harken gear on the foredeck, it was all very easy. Although this yacht is not cheap, the standard of hardware, quality of fittings and quite superb joinery work makes her good value. I was constantly impressed with the depth of thought and planning that had gone into the design and construction by the experienced and dedicated team that had constructed her. Here is a luxury yacht that fulfils her brief, and then some. She represents the very best in boat building skills from English yards and deserves to succeed. Perhaps it is fitting that although this is the first Premier yacht to be launched, already a 41ft version is in build, and there is serious interest in another 45ft and a 55ft version! The fortunate owners will have distinctive, seaworthy yachts that they can be proud of in any company.

Cost: Price, including the comprehensive fit out is approximately £280,000 (1992).

Footnote

The choice of steel as a building material for long distance cruising, particularly in coral waters is very attractive. The strength of the hull, in areas where an accidental stranding or collision with heavy debris cannot be ruled out, is of paramount importance. It is no coincidence that many respected and experienced ocean cruising yachtsmen are now choosing steel for serious offshore work. In the past, corrosion was always the problem, but now, with new, successful anti- corrosion lessons learnt in the harsh North Sea environments by the oil industry being applied to yacht construction, corrosion in steel has been controlled. The best yachts built by specialist companies such as Premier Yachts Ltd should have a life expectancy exceeding that of many other boat building materials.

Premier 45
LOA 47ft 8in
LW 36ft 3in
Beam 14ft
Draught 6ft 6in
Displacement 17.2 tons.

Premier Yachts Limited, Itchenore,
Chichester, West Sussex, PO20 7AY,
England. Telephone (0243) 512611.
Fax: (0243) 511473.

Arrow-head

A Dream Ocean Cruiser

Over the last forty years or so, having sailed and owned a multitude of yachts both mono and multihull has given me the opportunity of gradually honing my ideas concerning the ideal design for fast ocean cruising.

We have no ambitions to permanently live on board, but our dream yacht would have to be comfortable for extended cruising both in tropical and northern waters. She would need to be set up for easy, short handed sailing, yet be fast, stable, and capable of good passage times, without tiring her crew.

As one becomes older, ones priorities mature, and what seemed a wonderful idea in ones twenties, now is unthinkable. Moreover there has been tremendous developments in the dynamics of yacht hull design, sails and spars plus of course the almost giddy pace in the field of electronics, so it would be foolish not to take full advantage of this progress when contemplating a new design.

The main problem with age is that one's energy and strength diminish, also creature comforts start to loom large in one's priorities, so the emphasis regarding design changes somewhat. Labour-saving equipment and systems become a necessity and the ideal size for a yacht starts to lengthen. The problem with a larger boat is that it needs big sails to make it go and heavy ground tackle to keep it secure, plus of course handling in confined areas becomes more complicated and difficult; all factors that are in direct conflict with sailing a large boat with a small crew.

Large however, can be beautiful, for size directly effects comfort and speed at sea (not necessarily seaworthiness). Also, although it is a pleasure at times to cruise just with one's wife, there are also times when friends and family come along, so the requirement is there to accommodate them all in some degree of comfort.

So what is the ideal size? For some years now we have managed a forty five foot boat very well, but feel an extra few feet would give us additional benefits, notably, faster passage times, greater stability at sea and more comfort in harbour. Therefore we consider a light displacement fifty foot yacht that has been designed for easy handling of the primary functions, would be about ideal.

The Designer
We have firm views on what would constitute our ideal cruiser and in particular how it should handle and behave when offshore.

Our prime requirements are for a boat

with excellent directional stability, stiffness, and a hull that has good resistance to rhythmic rolling. As we enjoy 'sailing' our boats, she should be responsive and fast on all points, particularly downwind. Having suffered in the past from some older designs that feature fine lines aft, with the heavy quarter waves and the difficult steering that goes with them, we wanted a designer that had studied the modern approach to seakindliness, also someone that was forward looking, yet not stultified by the familiar limitations that can plague modern production yacht design. For us the choice was easy, for we had sailed on some of his recent designs and have been mesmerized with their seaworthiness and speed. Warwick Collins is also the designer of the brilliant tandem keel, something that we would have specified anyway, so this was just another bonus.

From the outset Warwick has been a pleasure to work with and he immediately grasped what was wanted and interpreted the brief with sensitivity and flair. The resulting design will be fast and stable, her pedigree in terms of seaworthiness is assured, and I only hope her owners will do her justice.

Hull

The design is typical Collins, with a fine entry and deep forefoot for good windward performance without slamming. The yacht has been given wide aftersections for stability and speed off the wind.

Displacement is on the light side of moderate, but she should be a reasonable load carrier. She has been designed for timber construction using the well proven West Epoxy system, so maintenance costs will be low, with no worries concerning rot or teredo worm attack.

A strong watertight bulkhead is positioned in the bow in case of a hull fracture, due to colliding with a large floating object when offshore.

Cockpit

In warm climates one tends to live in the cockpit so we specified one with a large area, and four big drains. The cockpit has been designed for sailing efficiency, with correct angles to the backrests and padded helmsman's positions for comfort and enjoyment when steering the yacht. The side benches are over six and a half feet long in order that they can be used for sleeping on if required. A wide smoked laminated glass windscreen keeps the cockpit dry, and with the hood erected it will be a snug place when going to windward. There is also provision for a folding Bimini to shade the occupants from the hot sun of the tropics.

The Rig and Sails

A high aspect ratio, sloop headed ketch rig of generous area was chosen for good performance, both on and off the wind. There are two headsails which are managed independently. The after sail is the one used under all normal circumstances, with the forward one being rolled out for down wind sailing. Together they will give approximately eighty percent of the area of a spinnaker, but without the handling problems. Both foresails are set on roller reefing systems and have independent poles attached to the mast. Mainsail and mizzen handling and reefing lines all lead aft to the cockpit. All are powered by electricity.

The mainsail and mizzen sheets are positioned close to the helmsman, as are the genoa winches so the boat can be handled if need be, by just one person. Main and mizzen sails are fully battened, with lazy-jacks and jiffy reefing. The storm jib is set on the inner stay and the trysail has its own track on the mainmast; both have their own separate sheets permanently bent-on ready for use. The sail wardrobe has the addition of a high cut, large area cruising chute for very light conditions, plus a mizzen staysail for the days when the crew want a little extra fun.

The Deck

The deck and coachroof will be teak laid, with good areas for sunbathing. The coachroof is a low profile affair that will be finished in clear varnish. The cabins will have their own hatches to provide good ventilation and light. The forward and aft hatches will be of the Goiot type that hinge both fore and aft. The life-raft has its own locker aft; as do the gas bottles; incorporating over the side drainage. There is also a large lazarette positioned aft with built in compartments for the outboard motor, inflatable dinghy, generator and fenders. The tools have their own fitted box and the extra anchors, chains and warps are all locked off securely.

There is an elegant multi purpose stainless steel gallery on the aft deck. It will incorporate the sliding track for the mizzen sheet, davits for the dinghy and a small swivelling crane to hoist the outboard motor from its stowage in the lazarette.

Anchors

Two bower anchors are positioned in their own separate lockers on the bow. The main bower, a 65lb CQR has forty fathoms of calibrated chain led to the power windlass and stowed at the stemhead. The second bower, a 45lb Fortress with three fathoms of chain and thirty seven fathoms of nylon warp stowed on a drum, is in the after bow locker. A small folding anchor with nylon warp will have its own small locker aft in the transom for when a stern anchor is required. To back these up will be a 65lb (the heaviest anchor one man can handle in a dinghy) fisherman with three fathoms of chain and thirty five fathoms of nylon warp. This beast will be securely lashed in the lazarette.

Interior

The interior has been built around the concept of efficient sea berths for six crew when offshore, and incorporates two comfortable en-suite cabins when in harbour or when used for gentle cruising. This boat has been designed to be used and cruised hard on the trade-wind routes; where the main activity will be centred around sleeping, navigating and eating. Therefore all these functions are positioned in the boat where the motion is least.

All berths will of course be fitted with lee cloths and the two passage berths on the starboard side with permanent boards, rather in the manner of the pilot berths of old.

The aft cabin has its own deck access via a low profile stainless steel ladder. The galley has enormous stowage with everything including the crockery in fitted compartments out of sight. The galley floor will have strips of unvarnished teak let into its surface for secure footage in a seaway. The three burner gas oven and the working surfaces adjacent, will be gimbaled as one unit so giving secure put-down space when cooking or brewing up. There will be a large top opening fridge with six inches of foam insulation plus a small freezer unit. The central sinks will drain on either tack. A provision has been made for a slide out work bench that can accommodate a portable vice. This is positioned under the bunk surface of the forward passage berth.

As this will be a timber yacht some of the inside surfaces of the hull will be left varnished. In general the fabrics will be light and modern. The saloon has been designed as a lounging, relaxing area and made as comfortable as possible with deep contour seating covered in light cream Connolly hide. The saloon table will fold up to a small coffee table size when not used as a full dining table.

Both heads incorporate separate shower compartments. The heads will be manually operated for simplicity and reliability. There will only be two openings in the hull. One for all water in, the other for all water out. These openings will be fitted with three inch

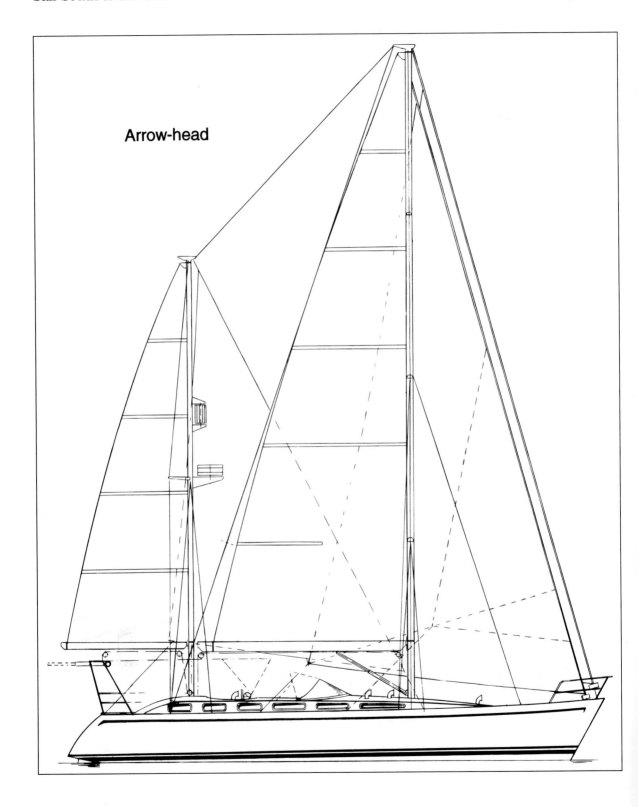

Arrow-head

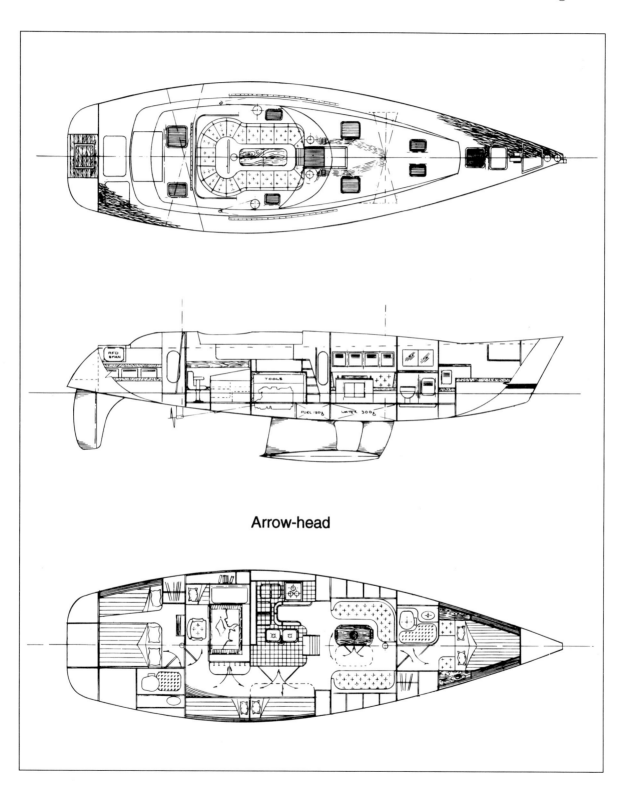

Arrow-head

Lloyds approved polypropylene seacocks, which in turn lead to a battery of small outlets/inlets, all fitted with individual shut-of valves.

Navigation

A great deal of thought has gone into this area of the yacht for it will be the nerve centre when on passage. Rather than incorporate it into the living area, as is usual, we have produced a separate navigation office or cabin that can be completely shut off from the other parts of the boat. It has a full size chart table with stowage for up to 100 charts in a special locker. Book stowage is generous and all electronic instruments can be housed on a special bulkhead that hinges down for maintenance. There is a purpose made stowage for the sextant. The navigator sits on a comfortable swivel seat with lateral supporting arm rests, so he is securely chocked in, whatever the weather. There is also a full length passage bunk for use when at sea.

Equipment

The engine will be approximately 100hp to give a relaxed cruising speed of around eight knots. Originally a shaft skeg was specified, but the design of the hull made this difficult, therefore a substantial A bracket will be fitted. The propeller will be a three bladed feathering type and fitted with a rope cutter. The engine will be flexibly mounted with heavy sound insulation carried out in the engine room.

The boat has been designed to be handled offshore by a minimum of crew so the sheet and halyard winches, plus the anchor winch will be electrically powered, with manual back up. The navigation equipment plus the auto pilot will be an integrated system from one manufacturer. The boat will also have radar and Navtex.

Tankage

Water is a vital commodity that has a direct effect on the quality of life at sea. Arrowhead will have 300 gallons of fresh water in two separate tanks.

Diesel fuel will also be in two separate tanks of 75 gallons each.

Plumbing and Power

Hot and cold water will be piped to the head compartments and galley plus cold water to an on-deck shower head.

A 24 volt system will be installed, with shore power and 220 volt main supply. Battery capacity will be four banks of 120 amps plus a separate 120 amp battery dedicated to engine starting, all stored low down in a semi sealed but ventilated compartment, safe from bilge water contamination.

Power generation will be from the main engine via a TWC unit, plus a small back up diesel generator, two solar panels and an Ampair water and wind generator. With all this equipment there should be little concern over power consumption whatever the conditions.

Performance

It is estimated that with the yachts slippery hull shape and efficient rig the boat will reach hull speed close-reaching in Force 4. Offwind she has been designed to maintain eight knots efficiently in trade-wind conditions without the use of a spinnaker, yet be stiff enough for windward sailing to remain unreefed up to Force 5. With a young energetic crew out for some fun, the boat will be capable of surfing downwind; yet always under control due to her clever hull and keel configurations.

In some ways she is an unusual yacht and at first glance not the sort of boat that one would expect for an ocean cruiser. It is only when looking at her in detail that her remarkable attributes become apparent. Warwick Collins has designed a boat that will be superbly seaworthy and capable of good passage times. She will be comfortable in harbour and at sea, yet be capable of being handled with a small crew under all conditions. She will also look stunning in any company, what more could one possible desire?

Arrow-head

The Designer's Comments

Mine has been the pleasant task to assist in the design of a dream yacht for Clifford Stillwell, including a number of features which reflect his experience of a variety of yachts he has owned or sailed on in the past.

Much of our discussion has revolved around his interest in a range of wood epoxy cruising yachts we have recently designed – in particular *Fondue, Sparkle* and *Baker's Lass*. In turn, I received from the author a great many interesting and cogent reasons for various features for his choice of interior layout, rig and other aspects of *Arrow-head*.

To many yacht design offices, racing yachts represent their primary interest and cruising yachts are largely a means of bringing in bread and butter commissions. If this is the case, I am an exception. I regard the cruising yacht as the primary form, and racing yachts as variations – quite interesting variations I must admit – on the cruising yacht.

Because the cruising yacht is not inhibited by rules, it is possible to design in a form which is easily handled and seaworthy, but potentially very fast. During the past two decades many fast cruisers have been subject to racing rules (I am thinking particularly of the IOR) which – almost criminally – have tended to penalise the attainment of optimum stability. The result has been offshore yachts which carry lead in the hull as opposed to the keel, and which require large strong crews to maintain seaworthiness in heavy weather.

The best modern cruising designs have evolved independently of IOR and often in wholly different directions. More beneficial influences have been the BOC Round-the-World Race and other long distance races such as OSTAR and TWOSTAR. In the evolution of cruising hulls, the modern shape of a narrow entry and powerful aft sections has created much better downwind characteristics than those associated with older hull shapes. The terrible rolling of old, deep, narrow hulled yachts has been alleviated. The modern shape is also much more resistant to pitch – not least because broad stern sections cannot be plunged up and down in water without generating considerable resistance.

Inventions such as the tandem keel have had a further beneficial effect by enhancing efficiency of keels at low draught. The end-plate on a typical tandem keel, containing approximately sixty percent of total volume and weight, ensures a very low centre of gravity and high righting moment for a given overall draught. But the shape of the keel further inhibits pitch, heave and roll, and reduces the strains on the helmsman in a seaway.

Modern rigs, with their enhanced systems of stowage and furling, have also contributed to ease of handling and safety. Like Clifford Stillwell, at this stage in the evolution of rigs I believe the fully battened mainsail, easily dropped and stowed within lazy jacks, combines high efficiency with ease of use.

It seems to me that *Arrow-head* is a product of all these considerations. I believe she will be comfortable and fast, with an easy motion in a seaway. Wood epoxy construction will ensure she will have a strong hull and a long life ahead of her.

Arrow-head

Principal Dimensions

LOA	50ft (15.24m)
LWL	44ft (13.40m)
BEAM	15ft (4.57m)
DRAUGHT	6ft 9in (2.06m)
DISP	12.8t (13,000kg)
BALLAST	5.9t (6000kg)

I	55ft 3in (16.85m)
J	15ft 11in (4.85m)
P	53ft (16.15m)
E	13.1ft (5.50m)
Py	34.1ft (10.40m)
Ey	11.6ft (3.50m)

Addendum

The following observation was made during a passage after the main text of this book had been completed.

The precise time to run off down wind depends on wind strength, sea state and the type and size of yacht involved. Usually conditions in deep water deteriorate sufficiently around Force 9 to contemplate this action although a yachts strategy might alter dramatically much earlier if she finds herself in shoaling waters or strong wind over tide conditions.

As mentioned I prefer to keep a yacht moving as fast as possible with the waves taken on the quarter, just so long as the yachts movement through the water does not disturb the seas astern and make them unstable. The main concern is to maintain good directional control and stop the boat from broaching as she is carried down the faces of the waves. Normally, in very heavy seas there is a moment on each wave when the wave lifts a yachts stern and carries her forward equalling the yachts speed. This can be a danger point for the rudder is stalled and is working in still water with a resultant loss of control. We have successfully overcome this problem when sailing in wind strengths of Force 10 and over by running the engine slowly ahead, not for propulsion, but to create some prop wash over the rudder, this gives it some bite for those few vital seconds before the yacht accelerates forward with the breaking crest.

Recommended Reading

Pilots

ROUTE - NORTHERN EUROPE TO
MEDITERRANEAN

Atlantic Spain & Portugal
By: RCC Pilotage Foundation,
Imray Laurie Norie & Wilson Ltd,

North Biscay Pilot
By: RCC Pilotage Foundation,
Adlard Coles Nautical

South Biscay Pilot
By: Robin Brandon,
Adlard Coles Nautical

Spain
By: Robin Brandon
3 vols
a. *Costa Del Sol and Blanca*
b. *Costas Del Azahar Dorada & Brava*
c. *Islas Balearics*
Imray Laurie Norie & Wilson Ltd

France
By: Robin Brandon
South France Pilot (West)
South France Pilot (East)
Imray Laurie Norie & Wilson Ltd

France & Corsica
By: Rod Heikell
Imray Laurie Norie & Wilson Ltd

Italian Waters Pilot
By: Rod Heikell
Imray Laurie Norie & Wilson Ltd

Adriatic Pilot
By: Trevor & Diana Thompson
Imray Laurie Norie & Wilson Ltd

Greek Waters Pilot
By: Rod Heikell
Imray Laurie Norie & Wilson Ltd

Turkish Water Pilot
By: Rod Heikell
Imray Laurie Norie & Wilson Ltd

Turkey & The Dodecanese Cruising Pilot
By: Robin Petherbridge
Adlard Coles Nautical

ROUTE - INLAND WATERWAYS

Cruising French Waterways
By: Hugh McKnight
Adlard Coles Nautical

Through the French Canals
By: Philip Bristow
Adlard Coles Nautical

French Waterway Guides
No's 1 to 41
Barnacle Marine Ltd.

ROUTE - NORTHERN EUROPE TO
CARIBBEAN

World Cruising Routes
By: Jimmy Cornell
Adlard Coles Nautical

The Atlantic Crossing Guide
By: Philip Allan & The RCC Pilotage
Foundation
Adlard Coles Nautical

The Atlantic Pilot Atlas
By: James Clarke
Adlard Coles Nautical

Ocean Passages for the World
The Hydrographic Office

Atlantic Islands
By: Anne Hamick & Nicholas Heath
RCC Pilotage Foundation
Imray Laurie Norie & Wilson Ltd

CARIBBEAN ISLANDS
Streets Guide to the Caribbean
By: Don Street
4 Vols
Imray Laurie Norie & Wilson Ltd

Lesser Antilles
By: RCC Pilotage Foundation
Imray Laurie Norie & Wilson Ltd

Cruising Guide to the Caribbean
By: Michael Marshall
Adlard Coles Nautical

Cruising Guide to the Virgin Islands
Sailors Guide to the Windward Islands
Cruising Guide to the Leeward Islands
By: Doyle
Barnacle Marine Ltd

NAVIGATION

Sight Reduction Tables for Air Navigation
Imray Laurie Norie & Wilson Ltd

Navigation for Yachtsman
By: Mary Blewitt
Adlard Coles Nautical

CHARTS

Admiralty Charts for World Coverage
Imray Charts for Northern Europe
Imray-Jolaine Charts for the Caribbean

GENERAL READING

Mediterranean Sailing
By: Rod Heikell
Adlard Coles Nautical

Mediterranean Cruising Handbook
By: Rod Heikell
Imray Laurie Norie & Wilson Ltd

This is Practical Weather Forecasting
By: Dieter Karnetzki
Adlard Coles Nautical

Cruising Under Sail
By: Eric Hiscock
Adlard Coles Nautical

World Cruising Handbook
By: Jimmy Cornell
Adlard Coles Nautical

Sell Up & Sail
By: Bill & Laurel Cooper
Adlard Coles Nautical

Care & Feeding of the Offshore Crew
By: Pardy
Barnacle Marine Ltd

Offshore Yachts: Characteristics
By: Rousmaniere
Barnacle Marine Ltd

Repairs at Sea
By: Nigel Calder
Adlard Coles Nautical

Boatowners Mechanical & Electrical Manual
By Nigel Calder
Adlard Coles Nautical

Seabirds
By: Peter Harrison
Adlard Coles Nautical

Heavy Weather Sailing
By: K. Adlard Coles
Adlard Coles Nautical

Some useful conversion tables

Metres to feet

m	ft
1	3.28
2	6.56
3	9.84
4	13.12
5	16.40
6	19.69
7	22.97
8	26.25
9	29.53
10	32.81
20	65.62
30	98.42
40	131.23
50	164.04
100	328.09

Centimetres to inches

cm	in
1	0.39
2	0.79
3	1.18
4	1.57
5	1.97
6	2.36
7	2.76
8	3.15
9	3.54
10	3.94
20	7.87
30	11.81
40	15.75
50	19.69
100	39.37

Kilogrammes to pounds

kg	lb
1	2.20
2	4.41
3	6.61
4	8.82
5	11.02
6	13.23
7	15.43
8	17.64
9	19.84
10	22.05
20	44.09
30	66.14
40	88.19
50	110.23
100	220.46

Litres to imperial gallons

litre	imp gal
1	0.22
2	0.44
3	0.66
4	0.88
5	1.10
6	1.32
7	1.54
8	1.76
9	1.98
10	2.20
20	4.40
30	6.60
40	8.80
50	11.10
75	16.50
100	22.00
200	44.00
500	110.00
1000	220.00

Kilometres to statute miles

km	m
1	0.62
5	3.11
8	4.97
10	6.00
20	12.00
30	19.00
40	25.00
50	31.00
75	47.00
100	62.00
250	155.00
500	311.00

Fathoms to feet to metres

fthm	ft	m
$\frac{1}{2}$	3	0.91
1	6	1.83
2	12	3.66
3	18	5.49
4	24	7.32
5	30	9.14
10	60	18.29
20	120	36.58
30	180	54.86

Cubic capacity

1 cu in =16.387cc
1 cu ft (1728 cu in) = 0.028m²
1 cu yard (27 cu ft) = 0.765m²
1 cu centimetre = 0.061 cu in
1 cu decimetre = 61.023 cu in
1 cu metre(1000cdm)=
35.315cuft
1 cu metre = 1.31 cu yd

Index

Some Useful Addresses

Anti-foul Paint
Awlgrip Gold,
Marineware Ltd, Unit 6, Cross House Centre, Cross House Road, Southampton, SO1 16Z, England. Tel. 0703 330208

Astro Course
Geoff Hayles, 6 Creek End, Emsworth, Hants. Tel. 0243 373756

Anti-slip Mats
Safestrip,
Copely Developments Ltd, Thurmaston Lane, Leicester, LE4 7HU. Tel. 765881

Anchors
Delta & C.Q.R.,
Simpson Lawrence Ltd, 218/228 Edminston Drive, Glasgow, G51 2YT. Tel. 041 427 5331/8
Fortress
Kelvin Hughes Ltd, 145 Minories, London, EC3 1NH. Tel. 0703 631286
Bruce
South Western Marine Factors Ltd., Pottery Road, Poole, Dorset, BH14 8RE. Tel.0202 745414

Anchor Chain Hook
Sinclair Everitt, 24 Anthony's Avenue, Lilliput, Poole, Dorset. Tel. 0202 707948

Folding Bicycles
Strida Ltd, Northway House, Northway, Cirencester, Glos. Tel. 0285 650333
Brompton Bicycles Ltd, The Arches, 2A London Road, Brentford, Middlesex, TW8 8JW. Tel. 01 847 0822

Nickel Alkaline Batteries
M.J. Furness, Schent Electronics Ltd, Unit14, Hamilton Way, Gore Road, New Milton, Hants. Tel. 0425 617976

Corrosion Protection for Steel Hulls
Copac System, Electrocataytic Ltd, Norman Way, Severn Bridge Industrial Estate, Portskewett, Newport, Gwent, NP6 4YN. Tel. 0291 423833

Diesel Fuel Additives
Aquasolve, Yacht Speed Enterprises Ltd, 77 Royal Hospital Road, London, SW3 4HN. Tel. 071 352 2830

Fuel Conditioning Unit
Powerplus Marine Ltd, Smalllock House, Crownhill, Plymouth, Devon. Tel. 0752 776700

Yacht Designer
Warwick Collins, 23 Kensington Park, Pennington, Lymington, Hants. SO41 8ES. Tel. 0590 679088

Diving Equipment
Sweeba Ltd, Southampton. Tel. 0489 581755

Dehydrated Food
Maid International Supply Company, PO Box 189, Granary House, St Peter Port, Guernsey, Channel Islands.

Hatches
Goiot Hatches, Montague Smith Ltd, 8 St Michael's Square, Southampton, SO9 4UY. Tel.0703 224667

Lubricant Spray
Gleit Technic Klaus Wietle GMB H, Ottostrause 182, 4100 Duisburg 17, Germany.

Marine Toilet Pump Adaptor
MIVA Products, PO Box 197, Southend-on-Sea, SS1 3EN. Tel. 0702 582754

MOB Equipment
Jon Buoy, Ocean Safety Ltd, 3 Crescent Stables, 139 Upper Richmond Road, London, SW15 2TX. Tel. 081 780 0111

Propeller Protectors
Ambassador Marine Ltd, 252 Hursley, Winchester, Hants, SO21 2JJ. Tel. 0962 75405
Harold Hayles Ltd., Yarmouth, Isle of Wight, PO41 0RS. Tel. 0983 760373

Refrigeration Equipment
Aqua Marine Ltd,216 Fair Oak Road, Bishopstoke, Eastleigh, Hants, SO5 6NJ.
Tel. 0703 694949
LV Motors Ltd., 1 Royston Road, Baldock, Herts, SG7 6NT, Herts. Tel. 0462 896095
Penguin Engineering Ltd, South Parade, Hayling Island, Hants. Tel. 0705 465607
South Western Marine Factors Ltd., Pottery Road, Poole, Dorset. Tel. 0202 745414

Rigging Testing
Maidsure Services, 9 Southern Road, West End, Southampton, Hants. Tel. 0703 472422

Shaft Seals
P.P.S. Shaft Seal, Fairways Marine Engineers, Downs Head, Maldon, Essex, CM9 7HU.
Tel. 0621 852866/859424
The Deep Sea Seal, Halyard Marine Ltd, 2 Portsmouth Centre, Quartremaine Road, Portsmouth Airport, Hants, PO3 5QT. Tel. 0705 671 641

Sea Anchor, Emergency Steering
Attenborough Sea Drogue, Fallowfield House, Puttenham, Guildford, Surrey, GU3 1AH.

Seals
Weather Tight Seals Ltd, 88 Upper Northam Road, Hedge End, Southampton, Hants, SO3 4EB.

Solar Panels
B.P. Solar International Ltd, 2 Aylesbury Vale Industrial Park, Stocklake, Aylesbury, HP20 1DQ. Tel. 0296 437555
Lumic Ltd, PO Box 416, Poole, Dorset, BH12 3LZ. Tel. 0202 749994
E.C. Smith Ltd, Units H & J, Kingsway Industrial Estate, Luton, Beds. Tel. 0582 29721
Marlec Ltd, Solarex Panels, Unit K, Cavendish Courtyard, Sallow Road, Corby, Northamptonshire, NN17 1DZ.Te. 0536 201588

Wind Vanes
Hydrovane, 117 Bramcote Lane, Chilwell, Nottingham, NG9 4EU. Tel. 0602 256181
Fleming, Navigair Ltd., Bridge Road, Swanwick, Southampton, Hants. Tel. 0489 885770
Sailomat, PO Box 10123, S-10055, Stockholm, Sweden.
Levanter & Sirius, Levanter Marine Equipment, 4 Gandish Road, East Bergholt, Suffolk. Tel. 0206 298242

Autosteer
Clearway Design, Kernick Industrial Estate, Penryn, Cornwall. Tel. 0326 76048